SpringerBriefs in Computer Science

For further volumes:
http://www.springer.com/series/10028

Robson L. F. Cordeiro
Christos Faloutsos · Caetano Traina Júnior

Data Mining in Large Sets of Complex Data

 Springer

Robson L. F. Cordeiro
Computer Science Department (ICMC)
University of São Paulo
São Carlos, SP
Brazil

Caetano Traina Júnior
Computer Science Department (ICMC)
University of São Paulo
São Carlos, SP
Brazil

Christos Faloutsos
Department of Computer Science
Carnegie Mellon University
Pittsburgh, PA
USA

ISSN 2191-5768   ISSN 2191-5776 (electronic)
ISBN 978-1-4471-4889-0   ISBN 978-1-4471-4890-6 (eBook)
DOI 10.1007/978-1-4471-4890-6
Springer London Heidelberg New York Dordrecht

Library of Congress Control Number: 2012954371

Printed on acid-free paper

Springer is part of Springer Science+Business Media (www.springer.com)

Preface

Both the amount and the complexity of the data gathered by current scientific and productive enterprises are increasing at an exponential rate, in the most diverse knowledge areas, such as biology, physics, medicine, astronomy, climate forecasting, etc. To find patterns and trends in these data is increasingly important and challenging for decision making. As a consequence, the analysis and management of *Big Data* is currently a central challenge in Computer Science, especially with regards to complex datasets. Finding clusters in large complex datasets is one of the most important tasks in data analyses. For example, given a satellite image database containing several tens of Terabytes, how can we find regions aiming at identifying native rainforests, deforestation, or reforestation? Can it be made automatically? Based on the results of the work discussed in this book, the answers to both questions are a sound "yes", and the results can be obtained in just minutes. In fact, results that used to require days or weeks of hard work from human specialists can now be obtained in minutes with high precision.

Clustering complex data is a computationally expensive task, and the best existing algorithms have a super-linear complexity regarding both the data set cardinality (number of data elements) and dimensionality (number of attributes describing each element). Therefore, those algorithms do not scale well, precluding being efficient to process large data sets. Focused on the analysis of *Big Data*, this book discusses new algorithms created to perform clustering in moderate-to-high dimensional data involving many billions of elements stored in several Terabytes of data, such as features extracted from large sets of complex objects, but that can nonetheless be quickly executed, in just a few minutes.

To achieve that performance, it was taken into consideration that high-dimensional data have the clusters bounded to a few dimensions each, thus existing only in subspaces of the original high-dimensional space, although each cluster can have correlations among dimensions distinct from those dimensions correlated in the other clusters. The novel techniques were developed to perform both hard and soft clustering (that is, assuming that each element can participate in just one or in several clusters that overlap in the space) that can be executed by serial or by parallel processing. Moreover, their applications are shown in several practical test cases.

Distinctly from most of the existing algorithms (and from all of the fastest ones), the new clustering techniques do not require the previous definition of the number of expected clusters, rather, it is inferred from the data and returned to the user. Besides, due to the assumption that each cluster exists because of correlations existing in a subset of the space dimensions, the new techniques not only find clusters with high quality and speed, but also spot the most significant dimensions for each cluster, a benefit that the previous algorithms only achieve at the expenses of costly processing.

The methodology to develop the techniques discussed in this book was based on the extension of hierarchical data structures, multidimensional multi-scaling analysis of the spatial data distribution based on a convolution process using Laplacian filters, on the evaluation of alternative cluster entropies, and on new cost functions that enable to evaluate the best strategies before executing them, allowing to perform a dynamic dataflow optimization of the parallel processing.

The new algorithms were compared with at least nine of the most efficient existing ones, and it was shown that the performance improvement is over at least one magnitude order, although always having its quality equivalent to the best achieved by the competing techniques. In extreme situations, it took just *two seconds* to obtain clusters from real data that the best competing techniques required *two days*, with equivalent accuracy. In one of the real cases evaluated, the new techniques described were able to find correct tags for every image from a data set containing several tens of thousands of images, performing soft clustering (thus assigning one or more tags to each image), using as guidelines the labeling performed by a user in not more than five images for each tag (that is, in at most 0.001 % of the image set). Experiments reported in the book show that the novel techniques achieved excellent results in real data from high impact applications, such as breast cancer diagnosis, region classification in satellite images, assistance to climate change forecast, recommendation systems for the Web, and social networks.

In summary, the work described here takes steps forward from traditional data mining (especially for clustering) by considering large, complex data sets. Note that, usually, current works focus on one aspect, either size or data complexity. The work described in this book considers both: it enables mining complex data from high impact applications; the data are large in the Terabyte-scale, not in Giga as usual; and very accurate results are found in just minutes. Thus, it provides a crucial and well-timed contribution for allowing the creation of *real time* applications that deal with *Big Data of high complexity* in which mining on the fly can make an immeasurable difference, like in cancer diagnosis or deforestation detection.

São Carlos, October 2012   Robson L. F. Cordeiro
                                                               Christos Faloutsos
                                                               Caetano Traina Júnior

Acknowledgments

This material is based upon work supported by FAPESP (São Paulo State Research Foundation), CAPES (Brazilian Coordination for Improvement of Higher Level Personnel), CNPq (Brazilian National Council for Supporting Research), Microsoft Research, and the National Science Foundation under Grant No. IIS1017415. Research was sponsored by the Defense Threat Reduction Agency and was accomplished under contract No. HDTRA1-10-1-0120, and by the Army Research Laboratory and was accomplished under Cooperative Agreement Number W911NF-09-2-0053. The views and conclusions contained in this document are those of the authors and should not be interpreted as representing the official policies, either expressed or implied, of the Army Research Laboratory or the U.S. Government. The U.S. Government is authorized to reproduce and distribute reprints for Government purposes notwithstanding any copyright notation here on. Any opinions, findings, and conclusions or recommendations expressed in this material are those of the authors and do not necessarily reflect the views of the National Science Foundation, or other funding parties.

The authors also thank the following collaborators for providing valuable support to the development of this work: Agma J. M. Traina, Fan Guo, U. Kang, Julio López, Donna S. Haverkamp, James H. Horne, Ellen K. Hughes, Gunhee Kim, Mirella M. Moro, Carlos A. Heuser and João Eduardo Ferreira.

Contents

Chapter 1
Introduction

Abstract This chapter presents an overview of the book. It contains brief descriptions of the facts that motivated the work, besides the corresponding problem definition, main objectives and central contributions. The following sections detail each one of these topics.

Keywords Knowledge discovery in databases · Data mining · Clustering · Labeling · Summarization · Big data · Complex data · Linear or quasi-linear complexity · Terabyte-scale data analysis

1.1 Motivation

The information generated or collected in digital formats for various application areas is growing not only in the number of objects and attributes, but also in the complexity of the attributes that describe each object [5, 7, 9–13]. This scenario has prompted the development of techniques and tools aimed at intelligently and automatically assist humans to analyze, to understand and to extract knowledge from raw data [5, 8, 13], molding the research area of *Knowledge Discovery in Databases (KDD)*.

The increasing amount of data makes the KDD tasks especially interesting, since they allow the data to be considered as useful resources in the decision-making processes of the organizations that own them, instead of being left unused in disks of computers, stored to never be accessed, such as real 'tombs of data' [6]. On the other hand, the increasing complexity of the data creates several challenges to the researchers, provided that most of the existing techniques are not appropriate to analyze complex data, such as images, audio, graphs and long texts. Common knowledge discovery tasks are clustering, classification and labeling, identifying measurement errors and outliers, inferring association rules and missing data, and dimensionality reduction.

R. L. F. Cordeiro et al., *Data Mining in Large Sets of Complex Data*,
SpringerBriefs in Computer Science, DOI: 10.1007/978-1-4471-4890-6_1,
© The Author(s) 2013

1.2 Problem Definition and Main Objectives

The knowledge discovery from data is a complex process that involves high computational costs. The complexity stems from a variety of tasks that can be performed to analyze the data and from the existence of several alternative ways to perform each task. For example, the properties of the various attributes used to describe each data object, such as the fact that they are categorical or continuous, the cardinality of the domains, and the correlations that may exist between different attributes, etc., they all make some techniques more suitable or prevent the use of others. Thus, the analyst must face a wide range of options, leading to a high complexity in the task of choosing appropriate mining strategies to be used for each case.

The high computational cost comes from the need to explore several data elements in different combinations to obtain the desired knowledge. Traditionally, the data to be analyzed are represented as numerical or categorical attributes in a table where each tuple describes an element in the set. The performance of the algorithms that implement the various tasks of data analysis commonly depend on the number of elements in the set, on the number of attributes in the table, and on the different ways in which both tuples and attributes interact with their peers. Most algorithms exhibit super-linear complexity regarding these factors, and thus, the computational cost increases fast with increasing amounts of data.

In the scope of complex data, such as images, audio, graphs and long texts, the discovery of knowledge commonly includes a preprocessing step, in which relevant features are extracted from each object. The features extracted must properly describe and identify each object, since they are actually used in search and comparison operations, instead of the complex object itself. Many features are commonly used to represent each object, and the resulting collection of features is named the feature vector.

The work described in this book focuses on the development of knowledge discovery techniques well-suited to analyze large collections of complex objects described *exclusively* by their feature vectors, especially for the task of clustering, but also tightly associated to other data mining tasks that are performed together. Thus, the following premise was explored:

Premise: *Although the same tasks commonly performed for traditional data are generally also necessary for the analysis of feature vectors extracted from complex objects, the complexity of the analysis and the computational cost associated increase significantly, preventing the use of most of the traditional techniques. Thus, knowledge discovery techniques well-suited to analyze large, complex datasets described exclusively by feature vectors need to be created. As specific properties of feature vectors can be taken into account, it is possible to reduce the complexity and the computational cost involved, which, in the case of high dimensional vectors, are naturally higher than those involved in traditional data.*

Therefore, the work is aimed at the development of techniques well-suited to analyze large collections of complex objects represented *exclusively* by their feature vectors, automatically extracted by preprocessing algorithms. Nevertheless, the techniques described can be applied to any kind of complex data, from which sets of numerical attributes of equal dimensionalities can be extracted. Thus, the analysis of multi-dimensional datasets is the scope of this book. The definitions related to this kind of data, which are used throughout the book, are presented as follows.

Definition 1.1 A **multi-dimensional dataset** $^d S = \{s_1, s_2, \ldots s_\eta\}$ is a set of η points in a d-dimensional space $^d\mathbb{S}$, $^d S \subset {}^d\mathbb{S}$, over the set of axes $E = \{e_1, e_2, \ldots e_d\}$, where d is the **dimensionality** of the dataset, and η is its **cardinality**.

Definition 1.2 A **dimension** (also called a feature or an attribute) $e_j \in E$ is an axis of the space where the dataset is embedded. Every axis related to a dataset must be orthogonal to the other axes.

Definition 1.3 A **point** $s_i \in {}^d S$ is a vector $s_i = (s_{i1}, s_{i2}, \ldots s_{id})$ that represents a data element in the space $^d\mathbb{S}$. Each value $s_{ij} \in s_i$ is a number in $\mathbb{R}$. Thus, the entire dataset is embedded in the d-dimensional hyper-cube $\mathbb{R}^d$.

1.3 Main Contributions

With regard to the task of clustering large sets of complex data, an analysis of the literature (see the upcoming Chap. 3) leads us to come to one main conclusion. In spite of the several qualities found in the existing works, to the best of our knowledge, there is no method published in the literature, and well-suited to look for clusters in sets of complex objects, that has *any* of the following desirable properties: (1) **linear or quasi-linear complexity** - to scale linearly or quasi-linearly in terms of memory requirement and execution time with regard to increasing numbers of points and axes, and; (2) **Terabyte-scale data analysis** - to be able to handle datasets of Terabyte-scale in feasible time. On the other hand, examples of applications with Terabytes of high dimensionality data abound: weather monitoring systems and climate change models, where we want to record wind speed, temperature, rain, humidity, pollutants, etc; social networks like Facebook TM, with millions of nodes, and several attributes per node (gender, age, number of friends, etc); astrophysics data, such as the Sloan Digital Sky Survey (SDSS), with billions of galaxies and attributes like red-shift, diameter, spectrum, etc. In fact, the analysis and management of *Big Data* is today one main concern of the Computer Science community, especially for high dimensional data. Thus, to overcome the two aforementioned limitations is extremely desirable nowadays.

The work described in this book focuses on overcoming both limitations. Specifically, it presents three novel, fast and scalable data mining algorithms well-suited to analyze large sets of complex data:

1. **The Method *Halite* for Correlation Clustering**: the algorithm *Halite* [2, 3] is a fast and scalable density-based clustering algorithm for multi-dimensional data able to analyze large collections of complex data elements. It creates a

multi-dimensional grid all over the data space and counts the number of points lying at each hyper-cubic cell provided by the grid. A hyper-quad-tree-like structure, called the Counting-tree, is used to store the counts. The tree is thereafter submitted to a filtering process able to identify regions that are, in a statistical sense, denser than its neighboring regions regarding at least one dimension, which leads to the final clustering result. The algorithm is fast and it has linear or quasi-linear time and space complexity regarding both the data size and the dimensionality. Therefore, *Halite* tackles the problem of **linear or quasi-linear complexity**.

2. **The Method *BoW* for Clustering Terabyte-scale Datasets**: the method *BoW* [4] focuses on the problem of finding clusters in Terabytes of moderate-to-high dimensionality data, such as features extracted from billions of complex data elements. In these cases, a serial processing strategy is usually impractical. Just to read a single Terabyte of data (at 5 GB/min on a single modern eSATA disk) one takes more than 3 hours. *BoW* explores parallelism and can treat as plug-in almost any of the serial clustering methods, including the algorithm *Halite*. The major research challenges addressed are (a) how to minimize the I/O cost, taking into account the *already existing* data partition (e.g., on disks), and (b) how to minimize the network cost among processing nodes. Either of them may become the bottleneck. The method *BoW* automatically spots the bottleneck and chooses a good strategy, one of them uses a novel *sampling-and-ignore* idea to reduce the network traffic. Specifically, *BoW* combines (a) potentially any serial algorithm used as a plug-in and (b) makes the plug-in run efficiently in parallel, by adaptively balancing the cost for disk accesses and network accesses, which allows *BoW* to achieve a very good tradeoff between these two possible bottlenecks. Therefore, *BoW* tackles the problem of **Terabyte-scale data analysis**.

3. **The Method *QMAS* for Labeling and Summarization**: the algorithm *QMAS* [1] uses the background knowledge of the clustering algorithms presented in this book to focus on two *distinct* data mining tasks—the tasks of labeling and summarizing large sets of complex data. Specifically, *QMAS* is a fast and scalable solution to two problems (a) *low-labor labeling*—given a large collection of complex objects, very few of which are labeled with keywords, find the most suitable labels for the remaining ones, and (b) *mining and attention routing*—in the same setting, find clusters, the top-N_O outlier objects, and the top-N_R representative objects. The algorithm is fast and it scales linearly with the data size, besides working even with tiny initial label sets.

These three algorithms were evaluated on *real, very large datasets* with up to *billions* of complex elements, and they always presented highly accurate results, being at least one order of magnitude faster than the fastest related works in almost all cases. The real life data used come from the following applications: automatic breast cancer diagnosis, satellite imagery analysis, and graph mining on a large web graph crawled by Yahoo![1] and also on the graph with all users and their connections from

---

[1] www.yahoo.com

the Twitter[2] social network. In extreme cases, the work presented in this book allowed to spot in only 2 seconds the clusters present in a large set of satellite images, while the related works took 2 days to perform the same task, achieving similar accuracy.

In summary, the work described in this book takes steps forward from traditional data mining (especially for clustering) by considering large, complex datasets. Note that, usually, current works focus in one aspect, either size or data complexity. The work described here considers both: it enables mining complex data from high impact applications, such as breast cancer diagnosis, region classification in satellite images, assistance to climate change forecast, recommendation systems for the Web and social networks; the data are large in the Terabyte-scale, not in Giga as usual; and very accurate results are found in just minutes. Thus, it provides a crucial and well timed contribution for allowing the creation of *real time* applications that deal with *Big Data of high complexity* in which mining on the fly can make an immeasurable difference, like in cancer diagnosis or deforestation detection.

1.4 Conclusions

This chapter provided an overview of the book with brief descriptions of the facts that motivated the work, besides its problem definition, main objectives and central contributions. The remaining chapters are structured as follows. In Chap. 2, an analysis of the literature is presented, including a description of relevant basic concepts. Some of the state-of-the-art works found in literature for the task of clustering multi-dimensional data with more than five or so dimensions are described in Chap. 3. Chapters 4–6 contain the central part of this book. They present the knowledge discovery techniques targeted by us, as well as experiments performed to evaluate these techniques. Finally, conclusions are given in Chap. 7.

References

1. Cordeiro, R.L.F., Guo, F., Haverkamp, D.S., Horne, J.H., Hughes, E.K., Kim, G., Traina, A.J.M., Traina Jr., C., Faloutsos, C.: Qmas: Querying, mining and summarization of multimodal databases. In: G.I. Webb, B. Liu, C. Zhang, D. Gunopulos, X. Wu (eds.) ICDM, pp. 785–790. IEEE Computer Society (2010).
2. Cordeiro, R.L.F., Traina, A.J.M., Faloutsos, C., Traina Jr., C.: Finding clusters in subspaces of very large, multi-dimensional datasets. In: F. Li, M.M. Moro, S. Ghandeharizadeh, J.R. Haritsa, G. Weikum, M.J. Carey, F. Casati, E.Y. Chang, I. Manolescu, S. Mehrotra, U. Dayal, V.J. Tsotras (eds.) ICDE, pp. 625–636. IEEE (2010).
3. Cordeiro, R.L.F., Traina, A.J.M., Faloutsos, C., Traina Jr., C.: Halite: Fast and scalable multiresolution local-correlation clustering. IEEE Transactions on Knowledge and Data Engineering 99(PrePrints) (2011). http://doi.ieeecomputersociety.org/10.1109/TKDE.2011.176. 16 pages

---

2 twitter.com

4. Cordeiro, R.L.F., Traina Jr., C., Traina, A.J.M., López, J., Kang, U., Faloutsos, C.: Clustering very large multi-dimensional datasets with mapreduce. In: C. Apté, J. Ghosh, P. Smyth (eds.) KDD, pp. 690–698. ACM (2011).
5. Fayyad, U.: A data miner's story - getting to know the grand challenges. In: Invited Innovation Talk, KDD (2007). Available at: http://videolectures.net/kdd07_fayyad_dms/
6. Fayyad, U.M.: Editorial. ACM SIGKDD Explorations **5**(2), 1–3 (2003)
7. Fayyad, U.M., Piatetsky-Shapiro, G., Smyth, P.: From data mining to knowledge discovery: An overview. In: Advances in Knowledge Discovery and Data Mining, pp. 1–34 (1996).
8. Fayyad, U.M., Uthurusamy, R.: Data mining and knowledge discovery in databases (introduction to the special section). Communications of the ACM **39**(11), 24–26 (1996)
9. Kanth, K.V.R., Agrawal, D., Singh, A.K.: Dimensionality reduction for similarity searching in dynamic databases. In: L.M. Haas, A. Tiwary (eds.) ACM SIGMOD International Conference on Management of Data, pp. 166–176. ACM Press, Seattle, Washington, USA (1998). Elaine Josiel.
10. Korn, F., Pagel, B.U., Faloutsos, C.: On the 'dimensionality curse' and the 'self-similarity blessing'. IEEE Transactions on Knowledge and Data Engineering (TKDE) 13(1), 96–111 (2001). doi: http://dx.doi.org/10.1109/69.908983
11. Kriegel, H.P., Kröger, P., Zimek, A.: Clustering high-dimensional data: A survey on subspace clustering, pattern-based clustering, and correlation clustering. ACM TKDD 3(1), 1–58 (2009). doi: http://doi.acm.org/10.1145/1497577.1497578
12. Pagel, B.U., Korn, F., Faloutsos, C.: Deflating the dimensionality curse using multiple fractal dimensions. In: IEEE International Conference on Data Engineering (ICDE), pp. 589–598. IEEE Computer Society, San Diego, CA (2000).
13. Sousa, E.P.M.: Identificação de correlações usando a teoria dos fractais. Ph.D. Dissertation, Computer Science Department—ICMC, University of São Paulo, USP, São Carlos, Brazil (2006) (in Portuguese).

Chapter 2
Related Work and Concepts

Abstract This chapter presents the main background knowledge relevant to the book. Sections 2.1 and 2.2 describe the areas of processing complex data and knowledge discovery in traditional databases. The task of clustering complex data is discussed in Sect. 2.3, while the task of labeling such kind of data is described in Sect. 2.4. Section 2.5 introduces the `MapReduce` framework, a promising tool for large scale data analysis, which has been proven to offer one valuable support to the execution of data mining algorithms in a parallel processing environment. Section 2.6 concludes the chapter.

Keywords Complex data · Moderate-to-high dimensionality data · Big data · Subspace clustering · Projected clustering · Correlation clustering · Labeling · MapReduce · Hadoop

2.1 Processing Complex Data

Database systems work efficiently with traditional numeric or textual data, but they usually do not provide complete support for complex data, such as images, videos, audio, graphs, long texts, fingerprints, geo-referenced data, among others. However, efficient methods for storing and retrieving complex data are increasingly needed [45]. Therefore, many researchers have been working to make database systems more suited to complex data processing and analysis.

The most common strategy is the manipulation of complex data based on features extracted automatically or semi-automatically from the data. This involves the application of techniques aimed at obtaining a set of features (the feature vector) to describe the complex element. Each feature is typically one value or one array of numerical values. The vector resulting from this process should properly describe the complex data, because the mining algorithms usually rely only on the features extracted to perform their tasks. It is common to find vectors containing hundreds or even thousands of features.

R. L. F. Cordeiro et al., *Data Mining in Large Sets of Complex Data*, 7
SpringerBriefs in Computer Science, DOI: 10.1007/978-1-4471-4890-6_2,
© The Author(s) 2013

For example, the extraction of features from images is usually based on the analysis of colors, textures, objects' shapes and their relationship. Due to its simplicity and low computational cost, the most used color descriptor is the histogram, which counts the numbers of pixels of each color in an image [43]. The color coherence vector [57], the color correlogram [33], the metric histogram [70] and the cells histogram [66] are other well-known color descriptors. Texture corresponds to the statistical distribution of how the color varies in the neighborhood of each pixel of the image. Texture analysis is not a trivial task and it usually leads to higher computational costs than the color analysis does. Statistical methods [24, 60] analyze properties, such as granularity, contrast and periodicity to differentiate textures, while syntactic methods [24] perform this task by identifying elements in the image and analyzing their spatial arrangements. Co-occurrence matrices [32], Gabor [20, 60] and wavelet transforms [14] are examples of such methods. The descriptors of shapes commonly have the highest computational costs compared to the other descriptors, thus, they are used mainly in specific applications [58]. There are two central techniques to detect shapes: the geometric methods of edge detection [11, 60] that analyze length, curvature and signature of the edges, and the scalar methods for region detection [63] that analyze the area, "eccentricity", and "rectangularity".

Besides the feature extraction process, there are still two main problems to be addressed in order to allow an efficient management of complex data. The first is the fact that the extractors usually generate many features (hundreds or even thousands). As described in the upcoming Sect. 2.2, it impairs the existing strategies for data storage and retrieval due to the "curse of dimensionality" [10, 37, 39, 49, 56]. Therefore, dimensionality reduction techniques are vital to the success of strategies for indexing, retrieving and analyzing complex data. The second problem stems from the need to compare complex data by similarity, because it usually does not make sense to compare them by equality, as it is commonly done for traditional data [74]. Moreover, the total ordering property does not hold among complex data elements— one can only say that two elements are equal or different, since there is no explicit rule to sort the elements. This fact distinguishes complex data elements even more from the traditional elements. The access methods based on the total ordering property do not support queries involving comparisons by similarity. Therefore, a new class of access methods was created, known as the Metric Access Methods (MAM), aimed at allowing searches by similarity. Examples of such methods are the *M-tree* [17], the *Slim-tree* [71], the *DF-tree* [72] and the *DBM-tree* [74], which are considered to be dynamic, as they allow data updates without the need to rebuild the structure.

The main principle for these techniques is the representation of the data in a metric space. The similarity between two elements is calculated by a distance function acting as a metric applied to the pair of elements in the same domain. The definition of one metric space is as follows.

Definition 2.1 One **metric space** is defined as a pair $\langle \mathbb{S}, m \rangle$, where $\mathbb{S}$ is the data domain and $m : \mathbb{S} \times \mathbb{S} \rightarrow \mathbb{R}^+$ a distance function acting as a metric. Given any $s_1, s_2, s_3 \in \mathbb{S}$, this function must respect the following properties: (1) symmetry, $m(s_1, s_2) = m(s_2, s_1)$; (2) non-negativity, $0 < m(s_1, s_2) < \infty, \forall\, s_1 \neq s_2$;

(3) identity, $m(s_1, s_1) = 0$; and (4) triangle inequality, $m(s_1, s_2) \leq m(s_1, s_3) + m(s_3, s_2)$, $\forall\, s_1, s_2, s_3 \in \mathbb{S}$.

One particular type of metric space is the d-dimensional space, for which a distance function is defined, denoted as $\langle {}^d\mathbb{S}, m \rangle$. This case is especially interesting to this book, since it allows posing similarity queries over complex objects represented by feature vectors of equal cardinality in a multi-dimensional dataset.

2.2 Knowledge Discovery in Traditional Data

The task of *Knowledge Discovery in Databases—KDD* is defined as follows.

> The nontrivial process of identifying valid, novel, potentially useful, and ultimately understandable patterns in data [28].

This process is commonly partitioned into three steps: *Preprocessing*, *Mining* and *Result Evaluation* [61]. To obtain high-level information from raw data, the data to be analyzed is usually represented as a set of points in a d-dimensional space for which a distance function acting as a metric is specified, as described in Sect. 2.1. The attributes of the data indicate dimensions and the data objects represent points in the space, while the similarity between pairs of objects is measured in terms of the respective distance function applied to the data space.

With the increasing quantity and complexity of the data generated or collected in digital systems, the task of *Preprocessing* has become essential to the whole KDD process [65]. In this step, the data are reduced and prepared by cleaning, integrating, selecting and transforming the objects to the subsequent mining step. A major problem to be minimized is the "curse of dimensionality" [10, 37, 39, 49, 56], a term referring to the fact that increasing the number of attributes in the objects quickly leads to significant degradation of the performance and also of the accuracy of existing techniques to access, store and process data. This occurs because data represented in high dimensional spaces tend to be extremely sparse and all the distances between any pair of points tend to be very similar, with respect to various distance functions and data distributions [10, 39, 56]. Dimensionality reduction is the most common technique applied to minimize this problem. It aims at obtaining a set of relevant and non-correlated attributes that allow representing the data in a space of lower dimensionality with minimum loss of information. The existing approaches are: feature selection, which discards among the original attributes the ones that contribute with less information to the data objects; and feature extraction, which creates a reduced set of new features, formed by linear combinations of the original attributes, able to represent the data with little loss of information [19].

The mining task is a major step in KDD. It involves the application of data mining algorithms chosen according to the user's objectives. Such tasks are classified as: predictive tasks, which seek a model to predict the value of an attribute based on the values of other attributes, by generalizing known examples, and descriptive tasks, which look for patterns that describe the intrinsic data behavior [64].

Classification is a major predictive mining task. It considers the existence of a training set with records classified according to the value of an attribute (target attribute or class) and a test set, in which the class of each record is unknown. The main goal is to predict the values of the target attribute (class) in the database records to be classified. The algorithms perform data classification by defining rules to describe correlations between the class attribute and the others. Examples of such algorithms are genetic algorithms [9, 82] and algorithms based on decision trees [13], on neural networks [25] or on the Bayes theorem (Bayesian classification) [81].

Clustering is an important descriptive task. Han and Kamber define it as follows.

> The process of grouping the data into classes or clusters, so that objects within a cluster have high similarity in comparison to one another but are very dissimilar to objects in other clusters [31].

Traditional clustering algorithms are commonly divided into: (1) hierarchical algorithms, which define a hierarchy between the clusters in a process that may be initiated with a single cluster, recursively partitioned in follow-up steps (*top–down*), or considering at first that each data object belongs to a distinct cluster, recursively merging the clusters latter (*bottom–up*); or (2) partitioning algorithms, which divide η objects into k clusters, $k \leq \eta$, such that each object belongs to at most one cluster and each cluster contains at least one object. Examples of well-known clustering methods are k-Means [42, 44, 67], k-Harmonic Means [80], DBSCAN [26] and STING [75].

The last step in the process of knowledge discovery is the result evaluation. At this stage, the patterns discovered in the previous step are interpreted and evaluated. If the patterns refer to satisfactory results (valid, novel, potentially useful and ultimately understandable), the knowledge is consolidated. Otherwise, the process returns to one previous stage to improve the results.

2.3 Clustering Complex Data

Complex data are usually represented by vectors with hundreds or even thousands of features in a multi-dimensional space, as described in Sect. 2.1. Each feature represents one dimension of the space. Due to the curse of dimensionality, traditional clustering methods, like k-Means [42, 44, 67], k-Harmonic Means [80], DBSCAN [26] and STING [75], are commonly inefficient and ineffective for these data [4–7]. The main factor that leads to their efficiency problem is that they often have super-linear computational complexity on both cardinality and dimensionality. Traditional methods are also often ineffective when applied to high dimensional data, since the data tend to be very sparse in the multi-dimensional space and the distances between any pair of points usually become very similar to each other, regarding several data distributions and distance functions [10, 39, 49, 56]. Thus, traditional methods do not solve the problem of clustering large, complex datasets [4–7, 23, 39, 49, 56, 73].

Dimensionality reduction methods minimize the effects of the dimensionality curse by finding a new set of orthogonal dimensions, of cardinality smaller than the original set's one, composed of non-correlated dimensions relevant to characterize the data. The elements of this set can be original dimensions, in the case of feature selection methods, or linear combinations of them, for feature extraction methods. Notice however that only global correlations are identified by such methods. In other words, dimensionality reduction methods look for correlations that occur for *all* dataset elements regarding a set of dimensions. Nevertheless, high dimensional data often present correlations local to subsets of the data elements and dimensions [23, 73]. Thus, distinct groups of data points correlated with different sets of dimensions may exist. Many of these correlations can also be non-linear. Therefore, it is clear that traditional dimensionality reduction techniques do not identify all possible correlations, as they evaluate correlations regarding the entire dataset, and thus, when used as a preprocessing step for traditional clustering, they do not solve the problem of clustering large, complex datasets [4–7, 23].

Since high dimensional data often present correlations local to subsets of elements and dimensions, the data are likely to present clusters that only exist in subspaces of the original data space. In other words, although high dimensional data usually do not present clusters in the space formed by all dimensions, the data tend to form clusters when the points are projected into subspaces generated by reduced sets of original dimensions or linear combinations of them. Moreover, different clusters may be formed in distinct subspaces. Several recent studies support this idea and a well-known survey on this area is presented at [39].

Figure 2.1 exemplifies the existence of clusters in subspaces of four 3-dimensional databases over the axes $E = \{x, y, z\}$. Figure 2.1a shows a 3-dimensional dataset projected onto axes x and y, while Fig. 2.1b shows the same dataset projected onto axes x and z. There exist two clusters in this data, C_1 and C_2. None of the clusters present a high density of points in the 3-dimensional space, but each cluster is a dense, elliptical object in one subspace. Thus, the clusters exist in subspaces only. Cluster C_1 exists in the subspace formed by axes x and z, while cluster C_2 exists in the subspace $\{x, y\}$. Besides elliptical clusters, real data may have clusters that assume *any* shape in their respective subspaces. The clusters must only be dense in that subspace. To illustrate this fact, we present 'star-shaped' and 'triangle-shaped' clusters in another dataset (Fig. 2.1c: x-y projection; Fig. 2.1d: x-z projection) and 'curved' clusters in one third example dataset (Fig. 2.1e: x-y projection; Fig. 2.1f: x-z projection).

Such clusters may also exist in subspaces formed by linear combinations of original axes. That is, clusters like the ones in our previous examples may be arbitrarily *rotated* in the space, thus not being aligned to the original axes. For example, Fig. 2.1g and h respectively plot x-y and x-z projections of one last 3-dimensional example dataset. Similarly to clusters C_1 and C_2, the clusters in this data are also dense, elliptical objects in subspaces, but, in this case, the subspaces are planes generated by linear combinations of the axes $\{x, y, z\}$. Traditional clustering is likely to fail when analyzing this data, as each cluster is spread over an axis (generated by linear combinations of the original axes) in the 3-dimensional space. Also, dimensionality

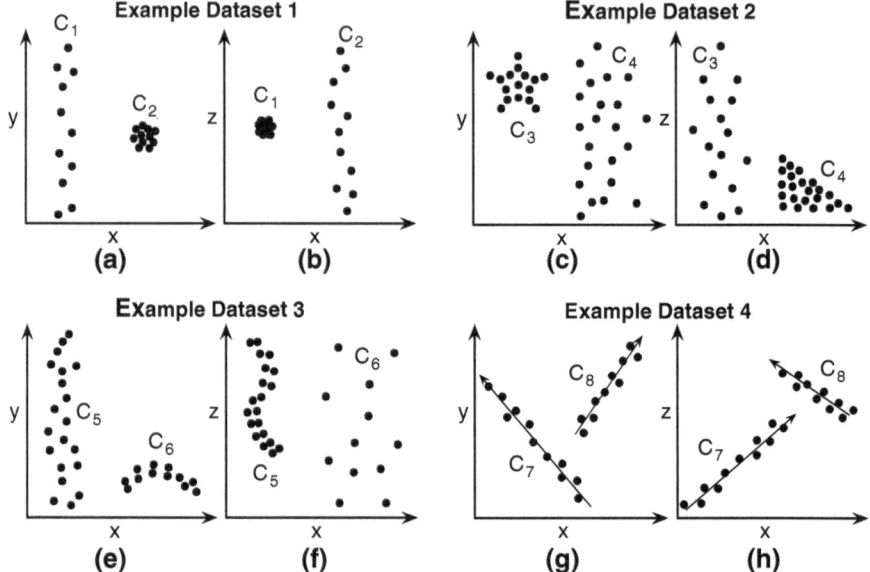

Fig. 2.1 x-y and x-z projections of four 3-dimensional datasets over axes $\{x, y, z\}$. From **a** to **f**: clusters in the subspaces $\{x, y\}$ and $\{x, z\}$. **g** and **h**: clusters in subspaces formed by linear combinations of $\{x, y, z\}$

reduction applied to the entire dataset does not help, as no 1- or 2-dimensional projection, axis aligned or not, keeps the clusters apart. This problem tends to be even worse in spaces of higher dimensionality. In fact, there is an interdependence between traditional clustering and dimensionality reduction that prevents them from solving the problem of clustering complex data. It is a fact that traditional clustering depends on a prior dimensionality reduction to analyze this kind of data. On the other hand, dimensionality reduction treats only global correlations, the ones that happen with regard to all data elements. Correlations local to subsets of the data cannot be identified without knowing the respective subsets. In other words, clustering complex data is not possible due to the high dimensionality, but the dimensionality reduction is incomplete without the prior knowledge of the data clusters where local correlations occur.

The most common strategy used to untangle this knotty problem is to unify both tasks, clustering and dimensionality reduction, creating a single task. Several methods have used this idea in order to look for clusters together with the subspaces where the clusters exist. According to a well-known survey [39], such methods differ from each other in two major aspects: (1) the search strategy used, which can be *top–down* or *bottom–up*, and (2) the characteristics of the clusters sought.

The *bottom–up* algorithms usually rely on a property named *downward closure* or *monotonicity*. This property, valid for some criteria of clustering characterization, warrants that: if there is at least one cluster in a d-dimensional dataset, at least one cluster will stand out when the data are projected into each of the possible subspaces formed by the original dimensions [7, 38, 39, 56]. As an example, this property is valid when the criterion used for clustering characterization is the *minimum density threshold*. By this criterion, one part of the data space contains a cluster if its density of points meet or exceed the minimum bound specified. The density of points never reduces when data are projected onto subspaces of the original space. Thus, if there is one part of a d-dimensional space whose density of points is sufficient to characterize a cluster, it is possible to affirm that at least one cluster will stand out (i.e., at least one space region will be dense enough) when the data are projected into any subspace formed by the original dimensions.

Based on a criterion of clustering characterization in which the *downward closure* property applies, *bottom–up* algorithms assume that: if a cluster exists in a space of high dimensionality, it has to exist or to be part of some cluster in all subspaces of lower dimensionality formed by original dimensions [7, 39, 56]. With that in mind, these methods start analyzing low dimensional subspaces, usually 1-dimensional ones, in order to identify the subspaces that contain clusters. The subspaces selected are then united in a recursive procedure that allows the identification of subspaces of higher dimensionality in which clusters also exist. Various techniques are used to spot and to prune subspaces without clusters so that they can be ignored in the process, minimizing the computational cost involved, but, in general, the main short-comings of *bottom–up* methods are: (1) they often have super-linear computational complexity or even exponential complexity regarding the dimensionality of the sub-spaces analyzed, and (2) fixed density thresholds are commonly used, assuming that clusters in high dimensional spaces are as dense as clusters in subspaces of smaller dimensionality, an assumption that is unlikely to be true in several cases.

The *top–down* algorithms assume that the analysis of the space with all dimensions can identify patterns that lead to the discovery of clusters existing in lower dimensional subspaces only [56]. This assumption is known in literature as the *locality assumption* [39]. After identifying a pattern, the distribution of points surrounding the pattern in the space with all dimensions is analyzed to define whether or not the pattern refers to a cluster and the subspace in which the possible cluster is better characterized. The main drawbacks of *top–down* algorithms are: (1) they often have super-linear computational complexity, though not exponential complexity, with regard to the data dimensionality, and; (2) there is no guarantee that the analysis of the data distribution in the space with all dimensions is always sufficient to identify clusters that exist in subspaces only.

Clustering algorithms for high dimensional data also differ from each other in the characteristics of the clusters that they look for. Some algorithms look for clusters that form a dataset partition, together with a set of outliers, while other algorithms partition the database regarding specific subspaces, so that the same data element can be part of two or more clusters that overlap with each other in the space with all dimensions, as long as the clusters are formed in different subspaces [56]. Finally, the subspaces analyzed by these methods may be limited to subsets of the original dimensions or may not be limited to them, thus including subspaces formed by linear combinations of the original dimensions [39].

Subspace clustering algorithms aim at analyzing projections of the dataset into spaces formed by subsets of the *original dimensions only*. Given a subspace and its corresponding data projection, these algorithms operate similarly to traditional clustering algorithms, partitioning the data into disjoint sets of elements, named **subspace** clusters, and one set of outliers. A data point can belong to more than one subspace cluster, as long as the clusters exist in different subspaces. Therefore, subspace clusters are *not necessarily disjoint*. Examples of subspace clustering algorithms are: CLIQUE [5, 7], ENCLUS [16], SUBCLU [34], FIRES [38], P3C [47, 48] and STATPC [46].

Projected clustering algorithms aim at partitioning the dataset into *disjoint* sets of elements, named **projected clusters**, and one set of outliers. A subspace formed by the *original dimensions* is assigned to each cluster, and the cluster elements are densely grouped, when projected into the respective subspace. Examples of projected clustering algorithms in literature are: PROCLUS [6], DOC/FASTDOC [59], PreDeCon [12], COSA [29], FINDIT [77], HARP [78], EPCH [51, 52], SSPC [79], P3C [47, 48] and LAC [8, 22, 23].

Correlation clustering algorithms aim at partitioning the database in a manner analogous to what occurs with projected clustering techniques—to identify *disjoint* clusters and one set of outliers. However, the clusters identified by such methods, named **correlation clusters**, are composed of densely grouped elements in subspaces formed by the original dimensions of the database, *or by their linear combinations*. Examples of correlation clustering algorithms found in literature are: ORCLUS [3, 4], 4C [12], CURLER [73], COPAC [1] and CASH [2]. Notice that, to the best of our knowledge, CURLER is the only method found in literature that spots clusters formed by non-linear, local correlations, besides the ones formed by linear, local-correlations.

2.4 Labeling Complex Data

In this section, we assume that the clustering algorithms aimed at analyzing complex data, such as the ones cited in the previous section, can also serve as a basis to perform one *distinct* data mining task—the task of labeling large sets of complex objects, as we discuss in the upcoming Chap. 6. For that reason, the following paragraphs introduce some background knowledge related to the task of labeling complex data, i.e., the task of analyzing a given collection of complex objects, in which a few objects are labeled with keywords, in order to spot appropriate labels for the remaining majority of unlabeled objects.

Specifically, the task of labeling is one predictive data mining task that considers the existence of a training set containing records labeled with keywords and a test set, in which the labels of each record are unknown. Its main goal is to assign appropriate keywords to the database records to be labeled. Unlike other data mining tasks, the task of labeling is not completely defined in the literature and slight conceptual divergences exist in the definitions provided by distinct authors, while some authors consider that labeling is one type of classification or use other names to refer to it (e.g., captioning). In this book, we consider that labeling refers to one generalized version of the classification task, in which the restriction of mandatorily assigning one and only one label to every data object does not exist. Specifically, we assume that labeling generalizes the task of classification with regard to at most three aspects: (1) it may consider that the dataset commonly contains objects that differ too much from the labeled training examples, which should be returned to the user as outliers that potentially deserve a new label of their own; (2) it may allow any object to receive more than one appropriate label; and (3) it may use hierarchies of labels, in a way that each object can be assigned to entire paths in the hierarchy, instead of being linked to individual labels only.

In the scope of complex data, labeling has been mostly applied to image datasets and also to sets of image regions segmented or arbitrarily extracted from larger images. There is an extensive body of work on the labeling of unlabeled regions extracted from partially labeled images in the computer vision field, such as image segmentation and region classification [41, 62, 69]. The Conditional Random Fields (CRF) and boosting approach [62] shows a competitive accuracy for multi-label labeling and segmentation, but it is relatively slow and requires many training examples. The KNN classifier [69] may be the fastest way for image region labeling, but it is not robust against outliers. Also, the Empirical Bayes approach [41] proposes to learn contextual information from unlabeled data. However, it may be difficult to learn the context from several types of complex data, such as from satellite images.

The principle Random Walks with Restarts (RWR) [68] has served as a basis to other labeling methods. The idea is to perform labeling by creating a graph to represent the input complex objects to be labeled, the given example labels and the similarities existing between the objects. Then, random walks in this graph allow spotting the most appropriate labels for the remaining unlabeled objects. In general, RWR consists into performing walks in a graph according to the following strategy: a random walker starts a walk from a vertex V of the graph, and, at each time step, the walker either goes back to the initial vertex V, with a user-defined probability c, or it goes to a randomly chosen vertex that shares an edge with the current vertex, with probability $1 - c$. The intuition is that this procedure provides an appropriate relevance score between any pair of graph nodes, since the steady state probability that a random walker will find itself in a vertex V', always restarting the walk from a vertex V, is a way to measure the closeness between V and V'. For labeling, RWR is commonly used to measure the closeness between the graph nodes that represent data objects and the ones that represent keywords.

GCap [54] is one of the most famous labeling methods based on random walks with restarts. It proposes a graph-based strategy for automatic image labeling that can also be applied to any set of multimedia objects. GCap represents images and label keywords by multiple layers of nodes in a graph and captures the similarities between pairs of images by creating edges to link the nodes that refer to similar images. The known labels become links between the respective images and keywords. This procedure creates a tri-partite graph that represents the input images and labels, besides the existing similarities between the images. Given an image node of interest, random walks with restarts are used to perform proximity queries in this graph, allowing GCap to automatically find the best annotation keyword for the respective image. Unfortunately, GCap remains rather inefficient, since it searches for the nearest neighbors of every image in the feature space to create edges between similar image nodes, and this operation is super-linear even with the speed up offered by approximate nearest-neighbor finding algorithms (e.g., the ANN Library [50]).

2.5 MapReduce

The large amounts of data collected by enterprises are accumulating data, and today it is already feasible to have *Terabyte-* or even *Petabyte-scale* datasets that must be submitted to data mining processes (e.g., Twitter crawl: > 12 TB, Yahoo! operational data: 5 *Petabytes* [27]), such as the processes of clustering and labeling complex data. However, the use of serial data mining algorithms, like the ones described in the previous sections, to analyze such huge amounts of data is clearly an impractical task. Just to read a single Terabyte of data (at 5 GB/min on a single modern eSATA disk) one takes more than 3 hours. Therefore, to improve the existing serial data mining methods in order to make them run efficiently in parallel is nowadays extremely desirable. With that in mind, this section describes the `MapReduce` framework, a promising tool for large-scale, parallel data analysis, which has been proving to offer one valuable support to the execution of data mining algorithms in a parallel processing environment.

`MapReduce` is a programming framework [21] fostered by Google[1] to process large-scale data in a massively parallel way. `MapReduce` has two major advantages: the programmer is oblivious of the details related to the data storage, distribution, replication, load balancing, etc.; and furthermore, it adopts the familiar concept of functional programming. The programmer needs to specify only two functions, a *map* and a *reduce*. The typical framework is as follows [40]: (a) the *map* stage passes over the input file and outputs (key, value) pairs; (b) the *shuffling* stage transfers the mappers output to the reducers based on the key, and; (c) the *reduce* stage processes the received pairs and outputs the final result. Due to its scalability, simplicity and the low cost to build large clouds of computers, `MapReduce` is a very promising tool for large scale, parallel data analysis, a fact that is already being reflected in the academia (e.g., [18, 35, 36, 55]).

`Hadoop` is the open source implementation of `MapReduce`. `Hadoop` provides the Hadoop Distributed File System (HDFS) [30], HBase [76], which is a way to efficiently store and handle semi-structured data as Google's BigTable storage system [15], and PIG, a high level language for data analysis [53].

---

[1] www.google.com

2.6 Conclusions

In this chapter we presented an overview of the basic concepts used in the book. We described the research areas of processing complex data and knowledge discovery in traditional databases, besides the main factors that distinguish the tasks of traditional clustering and clustering complex data. The task of labeling large sets of complex objects was also discussed. Finally, we introduced the `MapReduce` framework, a promising tool for large scale data analysis, which has been proving to offer one valuable support to the execution of data mining algorithms in a parallel processing environment. The next chapter describes some of the most relevant clustering methods available in literature for multi-dimensional data with more than five or so dimensions, which look for clusters formed by local correlations.

References

1. Achtert, E., Böhm, C., Kriegel, H.P., Kröger, P., Zimek, A.: Robust, complete, and efficient correlation clustering. SDM, USA, In (2007)
2. Achtert, E., Böhm, C., David, J., Kröger, P., Zimek, A.: Global correlation clustering based on the hough transform. Stat. Anal. Data Min. 1, 111–127 (2008). doi:10.1002/sam.v1:3
3. Aggarwal, C., Yu, P.: Redefining clustering for high-dimensional applications. IEEE TKDE 14(2), 210–225 (2002). doi:10.1109/69.991713
4. Aggarwal, C.C., Yu, P.S.: Finding generalized projected clusters in high dimensional spaces. SIGMOD Rec. 29(2), 70–81 (2000). doi:10.1145/335191.335383
5. Agrawal, R., Gehrke, J., Gunopulos, D., Raghavan, P.: Automatic subspace clustering of high dimensional data for data mining applications. SIGMOD Rec. 27(2), 94–105 (1998). doi:10.1145/276305.276314.
6. Aggarwal, C.C., Wolf, J.L., Yu, P.S., Procopiuc, C., Park, J.S.: Fast algorithms for projected clustering. SIGMOD Rec. 28(2), 61–72 (1999). doi:10.1145/304181.304188
7. Agrawal, R., Gehrke, J., Gunopulos, D., Raghavan, P.: Automatic subspace clustering of high dimensional data. Data Min. Knowl. Discov. 11(1), 5–33 (2005). doi:10.1007/s10618-005-1396-1
8. Al-Razgan, M., Domeniconi, C.: Weighted clustering ensembles. In: Ghosh, J., Lambert, D., Skillicorn, D.B., Srivastava, J. (eds.) SDM. SIAM (2006).
9. Ando, S., Iba, H.: Classification of gene expression profile using combinatory method of evolutionary computation and machine learning. Genet. Program Evolvable Mach. 5, 145–156 (2004). doi:10.1023/B:GENP.0000023685.83861.69
10. Beyer, K.S., Goldstein, J., Ramakrishnan, R., Shaft, U.: When is "nearest neighbor" meaningful? In: ICDT, pp. 217–235. UK (1999).
11. Blicher, A.P.: Edge detection and geometric methods in computer vision (differential topology, perception, artificial intelligence, low-level). Ph.D. thesis, University of California, Berkeley (1984). AAI8512758.
12. Bohm, C., Kailing, K., Kriegel, H.P., Kroger, P.: Density connected clustering with local subspace preferences. In: ICDM '04: Proceedings of the 4th IEEE International Conference on Data Mining, pp. 27–34. IEEE Computer Society, Washington, DC, USA (2004).
13. Breiman, L., Friedman, J.H., Olshen, R.A., Stone, C.J.: Classification and Regression Trees. Wadsworth, Belmont (1984)
14. Chan, T.F., Shen, J.: Image processing and analysis-variational, PDE, wavelet, and stochastic methods. SIAM (2005).

15. Chang, F., Dean, J., Ghemawat, S., Hsieh, W.C., Wallach, D.A., Burrows, M., Chandra, T., Fikes, A., Gruber, R.E.: Bigtable: a distributed storage system for structured data. In: USENIX'06. Berkeley, CA, USA (2006).

16. Cheng, C.H., Fu, A.W., Zhang, Y.: Entropy-based subspace clustering for mining numerical data. In: KDD, pp. 84–93. NY, USA (1999). doi:http://doi.acm.org/10.1145/312129.312199

17. Ciaccia, P., Patella, M., Zezula, P.: M-tree: an efficient access method for similarity search in metric spaces. In: The, VLDB Journal, pp. 426–435 (1997).

18. Cordeiro, R.L.F., Traina Jr., C., Traina, A.J.M., López, J., Kang, U., Faloutsos, C.: Clustering very large multi-dimensional datasets with mapreduce. In: Apté, C., Ghosh, J., Smyth, P. (eds.) KDD, pp. 690–698. ACM (2011).

19. Dash, M., Liu, H., Yao, J.: Dimensionality reduction for unsupervised data. In: Proceedings of the 9th IEEE International Conference on Tools with, Artificial Intelligence (ICTAI'97), pp. 532–539 (1997).

20. Daugman, J.G.: Uncertainty relation for resolution in space, spatial frequency, and orientation optimized by two-dimensional visual cortical filters. J. Opt. Soc. Am. A **2**, 1160–1169 (1985). doi:10.1364/JOSAA.2.001160

21. Dean, J., Ghemawat, S.: Mapreduce: simplified data processing on large clusters. OSDI (2004).

22. Domeniconi, C., Papadopoulos, D., Gunopulos, D., Ma, S.: Subspace clustering of high dimensional data. In: Berry, M.W., Dayal, U., Kamath, C., Skillicorn, D.B. (eds.) SDM (2004).

23. Domeniconi, C., Gunopulos, D., Ma, S., Yan, B., Al-Razgan, M., Papadopoulos, D.: Locally adaptive metrics for clustering high dimensional data. Data Min. Knowl. Discov. **14**(1), 63–97 (2007). doi:10.1007/s10618-006-0060-8

24. Duda, R., Hart, P., Stork, D.: Pattern Classification, 2nd edn. Wiley, New York (2001)

25. Duda, R.O., Hart, P.E., Stork, D.G.: Pattern Classification, 2nd edn. Wiley-Interscience, New York (2000)

26. Ester, M., Kriegel, H.P., Sander, J., Xu, X.: A density-based algorithm for discovering clusters in large spatial databases with noise. In: KDD, pp. 226–231 (1996).

27. Fayyad, U.: A data miner's story-getting to know the grand challenges. In: Invited Innovation Talk, KDD (2007). Slide 61. Available at: http://videolectures.net/kdd07_fayyad_dms/

28. Fayyad, U.M., Piatetsky-Shapiro, G., Smyth, P.: From data mining to knowledge discovery: an overview. In: Advances in Knowledge Discovery and Data Mining, pp. 1–34 (1996).

29. Friedman, J.H., Meulman, J.J.: Clustering objects on subsets of attributes (with discussion). J. Roy. Stat. Soc. B **66**(4), 815–849 (2004). doi:ideas.repec.org/a/bla/jorssb/v66y2004i4p815-849.html

30. Hadoop information. http://hadoop.apache.org/

31. Han, J., Kamber, M.: Data Mining: Concepts and Techniques, 2nd edn. Morgan Kaufmann, San Francisco (2006)

32. Haralick, R.M., Shanmugam, K., Dinstein, I.: Textural features for image classification. Syst. Man Cybern. IEEE Trans. **3**(6), 610–621 (1973). doi:10.1109/TSMC.1973.4309314

33. Huang, J., Kumar, S., Mitra, M., Zhu, W.J., Zabih, R.: Image indexing using color correlograms. In: Proceedings of 1997 IEEE Computer Society Conference on Computer Vision and, Pattern Recognition, pp. 762–768 (1997). doi:10.1109/CVPR.1997.609412.

34. Kailing, K., Kriegel, H.: Kroger. P, Density-connected subspace clustering for highdimensional data (2004).

35. Kang, U., Tsourakakis, C., Faloutsos, C.: Pegasus: a peta-scale graph mining system-implementation and observations. ICDM (2009).

36. Kang, U., Tsourakakis, C., Appel, A.P., Faloutsos, C., Leskovec., J.: Radius plots for mining tera-byte scale graphs: algorithms, patterns, and observations. SDM (2010).

37. Korn, F., Pagel, B.U., Faloutsos, C.: On the 'dimensionality curse' and the 'self-similarity blessing. IEEE Trans. Knowl. Data Eng. (TKDE) **13**(1), 96–111 (2001). doi:10.1109/69.908983

38. Kriegel, H.P., Kröger, P., Renz, M., Wurst, S.: A generic framework for efficient subspace clustering of high-dimensional data. In: ICDM, pp. 250–257. Washington, USA (2005). doi:http://dx.doi.org/10.1109/ICDM.2005.5

39. Kriegel, H.P., Kröger, P., Zimek, A.: Clustering high-dimensional data: a survey on subspace clustering, pattern-based clustering, and correlation clustering. ACM TKDD 3(1), 1–58 (2009). doi:10.1145/1497577.1497578
40. Lämmel, R.: Google's mapreduce programming model-revisited. Sci. Comput. Program. 70, 1–30 (2008)
41. Lazebnik, S., Raginsky, M.: An empirical bayes approach to contextual region classification. In: CVPR, pp. 2380–2387. IEEE (2009).
42. Lloyd, S.: Least squares quantization in pcm. Inf. Theory IEEE Trans. 28(2), 129–137 (1982). doi:10.1109/TIT.1982.1056489
43. Long, F., Zhang, H., Feng, D.D.: Fundamentals of content-based image retrieval. In: Multimedia Information Retrieval and Management. Springer (2002).
44. MacQueen, J.B.: Some methods for classification and analysis of multivariate observations. In: Cam, L.M.L., Neyman, J. (eds.) Proceedings of the 5th Berkeley Symposium on Mathematical Statistics and Probability, vol. 1, pp. 281–297. University of California Press (1967).
45. Mehrotra, S., Rui, Y., Chakrabarti, K., Ortega, M., Huang, T.S.: Multimedia analysis and retrieval system. In: Proceedings of 3rd International Workshop on Multimedia. Information Systems, pp. 25–27 (1997).
46. Moise, G., Sander, J.: Finding non-redundant, statistically significant regions in high dimensional data: a novel approach to projected and subspace clustering. In: KDD, pp. 533–541 (2008).
47. Moise, G., Sander, J., Ester, M.: P3C: a robust projected clustering algorithm. In: ICDM, pp. 414–425. IEEE Computer Society (2006).
48. Moise, G., Sander, J., Ester, M.: Robust projected clustering. Knowl. Inf. Syst 14(3), 273–298 (2008). doi:10.1007/s10115-007-0090-6
49. Moise, G., Zimek, A., Kröger, P., Kriegel, H.P., Sander, J.: Subspace and projected clustering: experimental evaluation and analysis. Knowl. Inf. Syst. 21(3), 299–326 (2009)
50. Mount, D.M., Arya, S.: Ann: a library for approximate nearest neighbor searching. http://www.cs.umd.edu/mount/ANN/
51. Ng, E.K.K., Fu, A.W.: Efficient algorithm for projected clustering. In: ICDE '02: Proceedings of the 18th International Conference on Data Engineering, p. 273. IEEE Computer Society, Washington, DC, USA (2002).
52. Ng, E.K.K., chee Fu, A.W., Wong, R.C.W.: Projective clustering by histograms. TKDE 17(3), 369–383 (2005). doi:10.1109/TKDE.2005.47.
53. Olston, C., Reed, B., Srivastava, U., Kumar, R., Tomkins, A.: Pig latin: a not-so-foreign language for data processing. In: SIGMOD '08, pp. 1099–1110 (2008).
54. Pan, J.Y., Yang, H.J., Faloutsos, C., Duygulu, P.: Gcap: graph-based automatic image captioning. In: CVPRW '04: Proceedings of the 2004 Conference on Computer Vision and Pattern Recognition, Workshop (CVPRW'04) vol. 9, p. 146 (2004).
55. Papadimitriou, S., Sun, J.: Disco: distributed co-clustering with map-reduce. ICDM (2008).
56. Parsons, L., Haque, E., Liu, H.: Subspace clustering for high dimensional data: a review. SIGKDD Explor. Newsl 6(1), 90–105 (2004). doi:10.1145/1007730.1007731
57. Pass, G., Zabih, R., Miller, J.: Comparing images using color coherence vectors. In: ACM Multimedia, pp. 65–73 (1996).
58. Pentland, A., Picard, R.W., Sclaroff, S.: Photobook: tools for content-based manipulation of image databases. In: Storage and Retrieval for Image and Video Databases (SPIE), pp. 34–47 (1994).
59. Procopiuc, C.M., Jones, M., Agarwal, P.K., Murali, T.M.: A monte carlo algorithm for fast projective clustering. In: SIGMOD, pp. 418–427. USA (2002). doi:http://doi.acm.org/10.1145/564691.564739
60. Rangayyan, R.M.: Biomedical Image Analysis. CRC Press, Boca Raton (2005)
61. Rezende, S.O.: Sistemas Inteligentes: Fundamentos e Aplicações. Ed , Manole Ltda (2002). (in Portuguese)
62. Shotton, J., Winn, J.M., Rother, C., Criminisi, A.: TextonBoost: joint appearance, shape and context modeling for multi-class object recognition and segmentation. In: Leonardis, A.,

Bischof, H., Pinz A. (eds.) ECCV (1), Lecture Notes in Computer Science, vol. 3951, pp. 1–15. Springer (2006).

63. Sonka, M., Hlavac, V., Boyle, R.: Image Processing: Analysis and Machine Vision, 2nd edn. Brooks/Cole Pub Co, Pacific Grove (1998)

64. Sousa, E.P.M.: Identificação de correlações usando a teoria dos fractais. Ph.D. Dissertation, Computer Science Department–ICMC, University of São Paulo-USP, São Carlos, Brazil (2006). (in Portuguese).

65. Sousa, E.P.: Caetano Traina, J., Traina, A.J., Wu, L., Faloutsos, C.: A fast and effective method to find correlations among attributes in databases. Data Min. Knowl. Discov. **14**(3), 367–407 (2007). doi:10.1007/s10618-006-0056-4

66. Stehling, R.O., Nascimento, M.A., Falcão, A.X.: Cell histograms versus color histograms for image representation and retrieval. Knowl. Inf. Syst. 5, 315–336 (2003). doi:10.1007/s10115-003-0084-y. http://portal.acm.org/citation.cfm?id=959128.959131

67. Steinhaus, H.: Sur la division des corp materiels en parties. Bull. Acad. Polon. Sci. 1, 801–804 (1956). (in French).

68. Tong, H., Faloutsos, C., Pan, J.Y.: Random walk with restart: fast solutions and applications. Knowl. Inf. Syst. 14, 327–346 (2008). doi:10.1007/s10115-007-0094-2. http://portal.acm.org/citation.cfm?id=1357641.1357646

69. Torralba, A.B., Fergus, R., Freeman, W.T.: 80 million tiny images: a large data set for nonparametric object and scene recognition. IEEE Trans. Pattern Anal. Mach. Intell. **30**(11), 1958–1970 (2008)

70. Traina, A.J.M., Traina, C., Bueno, J.M., Chino, F.J.T., Azevedo-Marques, P.: Efficient content-based image retrieval through metric histograms. World Wide Web 6, 157–185 (2003). doi:10.1023/A:1023670521530.

71. Traina Jr, C., Traina, A.J.M., Seeger, B., Faloutsos, C.: Slim-trees: high performance metric trees minimizing overlap between nodes. In: Zaniolo, C., Lockemann, P.C., Scholl, M.H., Grust, T. (eds.) International Conference on Extending Database Technology (EDBT). Lecture Notes in Computer Science, vol. 1777, pp. 51–65. Springer, Konstanz, Germany (2000).

72. Traina Jr., C., Traina, A.J.M., Santos Filho, R.F., Faloutsos, C.: How to improve the pruning ability of dynamic metric access methods. In: International Conference on Information and Knowledge Management (CIKM), pp. 219–226. ACM Press, McLean, VA, USA (2002).

73. Tung, A.K.H., Xu, X., Ooi, B.C.: Curler: finding and visualizing nonlinear correlation clusters. In: SIGMOD, pp. 467–478 (2005). doi:http://doi.acm.org/10.1145/1066157.1066211

74. Vieira, M.R., Traina Jr, C., Traina, A.J.M., Chino, F.J.T.: Dbm-tree: a dynamic metric access method sensitive to local density data. In: Lifschitz, S. (ed.) Brazilian Symposium on Databases (SBBD), vol. 1, pp. 33–47. SBC, Brasìlia, DF (2004)

75. Wang, W., Yang, J., Muntz, R.: Sting: a statistical information grid approach to spatial data mining. In: VLDB, pp. 186–195 (1997).

76. Wiki: http://wiki.apache.org/hadoop/hbase. Hadoop's Bigtable-like structure

77. Woo, K.G., Lee, J.H., Kim, M.H., Lee, Y.J.: Findit: a fast and intelligent subspace clustering algorithm using dimension voting. Inf. Softw. Technol. **46**(4), 255–271 (2004)

78. Yip, K.Y., Ng, M.K.: Harp: a practical projected clustering algorithm. IEEE Trans. on Knowl. Data Eng. 16(11), 1387–1397 (2004). doi:http://dx.doi.org/10.1109/TKDE.2004.74. Member-David W. Cheung

79. Yip, K.Y., Cheung, D.W., Ng, M.K.: On discovery of extremely low-dimensional clusters using semi-supervised projected clustering. In: ICDE, pp. 329–340. Washington, USA (2005). doi:http://dx.doi.org/10.1109/ICDE.2005.96

80. Zhang, B., Hsu, M., Dayal, U.: K-harmonic means-a spatial clustering algorithm with boosting. In: Roddick, J.F., Hornsby, K. (eds.) TSDM. Lecture Notes in Computer Science, vol. 2007, pp. 31–45. Springer (2000).

81. Zhang, H.: The optimality of naive Bayes. In: V. Barr, Z. Markov (eds.) FLAIRS Conference. AAAI Press (2004). http://www.cs.unb.ca/profs/hzhang/publications/FLAIRS04ZhangH.pdf

82. Zhou, C., Xiao, W., Tirpak, T.M., Nelson, P.C.: Evolving accurate and compact classification rules with gene expression programming. IEEE Trans. Evol. Comput. **7**(6), 519–531 (2003)

Chapter 3
Clustering Methods for Moderate-to-High Dimensionality Data

Abstract Traditional clustering methods are usually inefficient and ineffective over data with more than five or so dimensions. In Sect. 2.3 of the previous chapter, we discuss the main reasons that lead to this fact. It is also mentioned that the use of dimensionality reduction methods does not solve the problem, since it allows one to treat only the global correlations in the data. Correlations local to subsets of the data cannot be identified without the prior identification of the data clusters where they occur. Thus, algorithms that combine dimensionality reduction and clustering into a single task have been developed to look for clusters together with the subspaces of the original space where they exist. Some of these algorithms are briefly described in this chapter. Specifically, we first present a concise survey on the existing algorithms, and later we discuss three of the most relevant ones. Then, in order to help one to evaluate and to compare the algorithms, we conclude the chapter by presenting a table to link some of the most relevant techniques with the main desirable properties that any clustering technique for moderate-to-high dimensionality data should have. The general goal is to identify the main strategies already used to deal with the problem, besides the key limitations of the existing techniques.

Keywords Complex data · Moderate-to-high dimensionality data · Subspace clustering · Projected clustering · Correlation clustering · Linear or quasi-linear complexity · Terabyte-scale data analysis

3.1 Brief Survey

CLIQUE [7, 8] was probably the first technique aimed at finding clusters in subspaces of multi-dimensional data. It uses a bottom-up approach, dividing 1-dimensional data projections into a user-defined number of partitions and merging dense partitions to spot clusters in subspaces of higher dimensionality. CLIQUE scales exponentially on the cluster dimensionality and it relies on a fixed density threshold that assumes

R. L. F. Cordeiro et al., *Data Mining in Large Sets of Complex Data*,
SpringerBriefs in Computer Science, DOI: 10.1007/978-1-4471-4890-6_3,
© The Author(s) 2013

high-dimensional clusters to be as dense as low-dimensional ones, an assumption that is often unrealistic. Many works, such as ENCLUS [16], EPCH [28], P3C [26, 27], SUBCLU [24] and FIRES [22] improve the ideas of CLIQUE to reduce its drawbacks, but they are still typically super-linear in time or in space.

PROCLUS [5] introduced the top-down clustering strategy, assuming what is known in the literature as the *locality assumption*: the analysis of the space with all dimensions is sufficient to find patterns that lead to clusters that only exist in subspaces. PROCLUS is a k-medoid method that assigns to each medoid a subspace of the original axes. Each point is assigned to the closest medoid in its subspace. An iterative process analyzes the points distribution of each cluster in each axis and the axes in which the cluster is denser form its subspace. PROCLUS scales super-linearly on the number of points and axes and it only finds clusters in subspaces formed by the original axes. ORCLUS [4, 6] and CURLER [31] improve the ideas of PROCLUS to find arbitrarily oriented correlation clusters. They analyze each cluster's orientation, based on the data attributes eigenvector with the biggest eigenvalue, in an iterative process that merges close clusters with similar orientations. CURLER spots even non-linear, local correlations, but it has a quadratic time complexity regarding the number of clusters and their dimensionalities, and its complexity is cubic with respect to the data dimensionality. Other well-known, top-down methods are: DOC/FASTDOC [29], PkM [3], LAC [18], RIC [12], LWC/CLWC [17], PreDeCon [14], OCI [13], FPC/CFPC [33], COSA [20], HARP [32] and STATPC [25].

Density-based strategies have also been used for correlation clustering. 4C [15] finds arbitrarily oriented clusters, by extending each cluster from a seed as long as a density-criterion is fulfilled. Otherwise, it picks another seed, until all points are classified. The density criterion used is a minimal required number of points in the neighborhood of each point, and the neighborhood is defined with a distance function that uses as a basis the eigensystems of its input points. 4C tends to uncover clusters of a single, user-defined dimensionality at each run, but the algorithm COPAC [2] improves the ideas of 4C to fix the issue.

This section presented a brief survey on existing algorithms well-suited to analyze moderate-to-high dimensionality data. One detailed survey on this area is in [23]. The next sections concisely describe three of the most relevant algorithms.

3.2 CLIQUE

CLIQUE [7, 8] was probably the first method aimed at finding clusters in subspaces of multi-dimensional data. It proposes a *bottom-up* search strategy to identify *subspace clusters*. The process starts by analyzing the input data projected into the 1-dimensional subspaces formed by each of the original dimensions. One data scan is performed to partition each projection into ξ equal sized intervals, which enables the creation of histograms that represent the points distribution regarding each interval and dimension. An interval is considered to be a *dense unit* when the percentage of

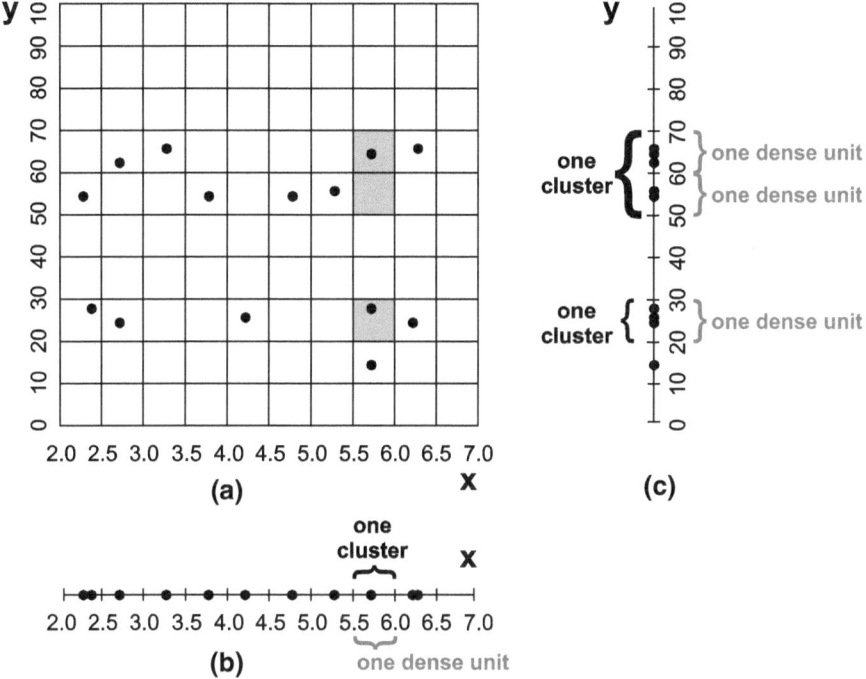

Fig. 3.1 Examples of 1-dimensional dense units and clusters existing in a toy database over the dimensions $A = \{x, y\}$

points inside it, with regard to the total number of points, is greater than or equal to a density threshold τ. The values of ξ and τ are user-defined input parameters.

Examples of dense units are illustrated in Fig. 3.1. Specifically, Figs. 3.1a, b and c respectively present an example dataset over the set of dimensions $A = \{x, y\}$ and its projections in the 1-dimensional subspaces formed by dimensions x and y. The database contains a total of 14 elements and each 1-dimensional projection is partitioned into $\xi = 10$ equal sized intervals. Given a density threshold $\tau = 20\%$ and one of these intervals, CLIQUE assumes that the interval is a dense unit if and only if it has at least 3 points. Consequently, after one data scan, the algorithm finds that there are three dense units in our toy dataset with regard to the projection in y and one dense unit in the projection in x. These are identified as $(20 \leq y < 30)$, $(50 \leq y < 60)$, $(60 \leq y < 70)$ and $(5.5 \leq x < 6.0)$, according to their respective ranges of values in y or in x.

Once the 1-dimensional dense units are identified, the process continues in order to find k-dimensional dense units, for $k > 1$. A pair of dense units in subspaces with $k - 1$ dimensions generates a candidate k-dimensional dense unit if and only if the respective subspaces share exactly $k - 2$ dimensions and both units are in the same positions regarding the shared dimensions. After identifying all candidate k-dimensional dense units, one data scan allows the algorithm to verify if the candidates are actually dense units. A candidate is said to be a dense unit if the number of points that fall into it is enough with regard to the threshold τ. The process continues recursively and it ends when, in an iteration: new dense units are not found, or; the space with all dimensions is analyzed.

In our running example from Fig. 3.1, the 1-dimensional dense units identified lead to the existence of three candidates, 2-dimensional dense units, filled in gray in Fig. 3.1a. These are described as $(20 \leq y < 30) \land (5.5 \leq x < 6.0)$, $(50 \leq y < 60) \land (5.5 \leq x < 6.0)$ and $(60 \leq y < 70) \land (5.5 \leq x < 6.0)$. However, after one data scan, it is possible to notice that none of the candidates has three or more points, which is the minimum number of points needed to spot a dense unit in this example. Thus, the recursive process is terminated.

CLIQUE also proposes an algorithm to prune subspaces in order to minimize its computational costs. It considers that the larger the sum of points in the dense units of a subspace analyzed, the more likely that this subspace will be useful in the next steps of the process. Thus, some subspaces analyzed by the method that have small sums of points in their dense units may be ignored in follow-up steps. For example, consider again our toy dataset from Fig. 3.1. The sums of points in the dense units of the 1-dimensional subspaces formed by dimensions x and y are 3 and 13 respectively. Thus, CLIQUE considers that, if needed, it is better to prune the subspace formed by dimension x than the one formed by y.

Pruning is performed as follows: at each iteration of the process, the dense units found are grouped according to their respective subspaces. The subspaces are then put into a list, sorted in descending order regarding the sum of points in their dense units. The principle *Minimum Description Length (MDL)* [21, 30] is then used to partition this list into two sublists: *interesting subspaces* and *uninteresting subspaces*. The MDL principle allows maximizing the homogeneity of the values in the sublists regarding the sum of points in the dense units of each subspace. After that, the dense units of the *uninteresting subspaces* are discarded and the recursive process continues, considering only the remaining dense units.

Once the recursive process is completed, some dense units identified are merged so that maximum sets of dense units adjacent in one subspace indicate clusters in that subspace. The problem of finding maximum sets of adjacent, dense units is equivalent to a well-known problem in the Graph Theory: the search for connected subgraphs. This is verified considering a graph whose vertices correspond to dense units in one subspace and edges connect two vertices related to adjacent, dense units, i.e., the ones that have a common face in the respective subspace. Thus, a *depth-first* search algorithm [9] is applied to find the maximum sets of adjacent, dense units defining the clusters and their corresponding subspaces. In our example dataset, the dense units found in Fig. 3.1c form two clusters in the subspace defined by dimension y. One cluster is related to the dense unit $(20 \leq y < 30)$, while the other refers to the pair of adjacent dense units $(50 \leq y < 60)$ and $(60 \leq y < 70)$. Finally, the subspace of dimension x contains a single cluster, represented by the dense unit $(5.5 \leq x < 6.0)$ in Fig. 3.1b, and no cluster exists in the 2-dimensional space.

The main shortcomings of CLIQUE are: (1) even using the pruning of subspaces proposed, the time complexity of the algorithm is still exponential with regard to the dimensionality of the clusters found; (2) due to the pruning technique used, there is no guarantee that all clusters will be found; (3) there is no policy suggested to define values appropriate to the parameters ξ and τ, and; (4) the density parameter τ assumes that clusters in subspaces of high dimensionality should be as dense as clusters in subspaces of lower dimensionality, one assumption often unrealistic.

3.3 LAC

LAC [10, 18, 19] is a k-means-based method aimed at clustering moderate-to-high dimensionality data. Given one user-defined number of clusters k, the algorithm partitions the dataset into k disjoint sets of elements, disregarding the possible existence of outliers. Its main difference from the traditional k-means method is to consider that any original dimension can be more relevant or less relevant than the other ones to characterize each cluster. The more the points of a cluster are densely grouped when projected into a dimension, the highest is the relevance of that dimension to the cluster. In this way, one dimension may be highly relevant to one cluster and, at the same time, it may have a lower relevance with regard to other clusters.

The final clustering result is a set of k data clusters and k weighting vectors. The weighting vectors represent the relevance of each original dimension with regard to each cluster found, in a way that the elements in each cluster are densely grouped into the space with all dimensions according to the L_2 distance function, weighted by the corresponding vector. Provided that LAC defines one dataset partition by only analyzing original dimensions, we consider that it finds clusters similar to *projected clusters*, besides the fact that LAC does not clearly define the subspaces where the clusters found exist.

Given a database S with N points on an Euclidean D-dimensional space, the algorithm computes a set of k centroids $\{c_1, c_2, \ldots c_k\}$, for $c_j \in \mathbb{R}^D$ and $1 \leq j \leq k$, together with a set of weighting vectors $\{w_1, w_2, \ldots w_k\}$, for $w_j \in \mathbb{R}^D$ and $1 \leq j \leq k$. The centroids and their weighting vectors define k clusters $\{S_1, S_2, \ldots S_k\}$ in a way that the clusters minimize the sum of the squared L_2 weighted distances between each data point and the centroid of its cluster, according to the corresponding weighting vector. The proposed procedure ensures that $S = S_1 \cup S_2 \cup \ldots S_k$ will always be true, and it also guarantees that, given any two clusters, S_a and S_b, the expression $S_a \cap S_b = \emptyset$ will always apply, for $1 \leq a \leq k$, $1 \leq b \leq k$ and $a \neq b$.

LAC uses a *top-down* clustering strategy that starts with the choice of k centroids. The first centroid is chosen at random. The second one maximizes the L_2 distance to the first centroid. The third centroid maximizes the L_2 distances to the two centroids chosen before, and so on. Equal weights are initially defined for all centroids and dimensions. Therefore, the initial dataset partition is found by assigning each point to the cluster of its nearest centroid, based on the unweighted L_2 distance function.

After the initial clusters are defined, LAC improves the centroids and the weighting vectors by minimizing the sum of the squared L_2 distances between each data point and the centroid of its cluster, weighted by the corresponding weighting vector.

At first, the weighting vectors are updated by analyzing the distribution of the points of each cluster projected into each original dimension individually. The more the points of a cluster are densely grouped when projected into a dimension, the biggest is the weight of that dimension to the cluster. During this process, LAC uses a user-defined parameter h, $0 \leq h \leq \infty$, to control how much the distribution of the values in each weighting vector will deviate from the uniform distribution. Setting $h = 0$ concentrates all the weight that refers to a cluster j on a single axis (the axis i in which the points of j projected into i are best clustered, compared to when these points are projected into each one of the other axes), whereas setting $h = \infty$ forces all axes to be given equal weights for cluster j, regardless of the data distribution. Values of h between 0 and ∞ lead to intermediate results.

Once the weighting vectors are updated for the first time, the data is partitioned again by assigning each point to the cluster of its closest centroid according to the L_2 distance function, weighted by the updated centroid's weighting vector. This process defines new centroids, and thus, it creates new clusters, whose point distribution must be analyzed for each dimension to update the weighting vectors one more time. It leads to a recursive process that stops when, after one iteration, none of the weighting vectors change.

The main shortcomings of LAC are: (1) the number of clusters k is user-defined; (2) LAC is non-deterministic, since the first centroid is randomly chosen at the beginning of the process; (3) although the authors present a formal proof of convergence for LAC, there is no guarantee that the convergence will occur in feasible time for all cases, and; (4) LAC disregards the possible existence of outliers in the data, which is considered to be a shortcoming, since existing methods for outlier detection do not treat the case where clusters exist in subspaces of the original space.

3.4 CURLER

CURLER [31] is a *correlation clustering* algorithm that spots clusters using a *top-down* strategy. To the best of our knowledge, it is the only method in literature that spots clusters formed by local, non-linear correlations, besides clusters formed by local, linear correlations. Although its positive aspects are relevant, CURLER cannot be fully automated, depending on a semi-automatic process in which the user visualizes unrefined clusters found by the algorithm to produce the final result.

Similarly to other correlation clustering algorithms, CURLER starts the clustering procedure by analyzing the space with all original axes to find tiny sets of points densely grouped, which are named microclusters. Then, some of the microclusters are merged to form correlation clusters. In general, microclusters that are close to each other in specific subspaces and have similar orientations should be merged, where the orientation of a microcluster is the eigenvector with the biggest eigenvalue found for the corresponding data objects. However, CURLER points out that the microclusters to be merged must be carefully identified, especially when analyzing data with clusters formed by non-linear, local correlations, since these clusters usually have one

Fig. 3.2 Examples of global
and local orientations in a toy
2-dimensional database

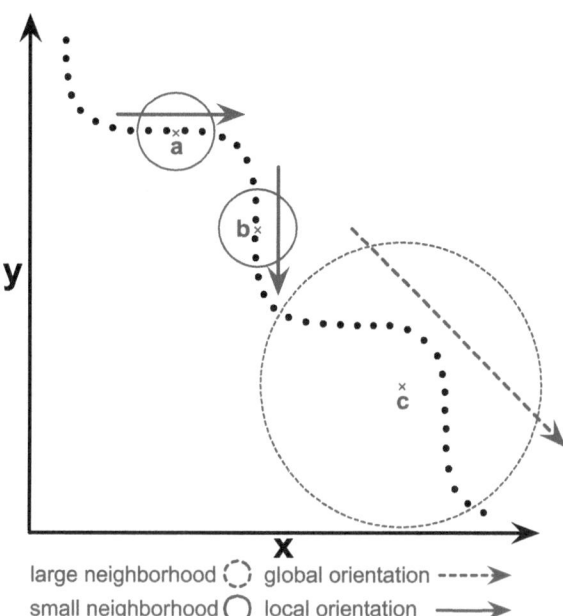

global orientation and several distinct local orientations. The global orientation is
the eigenvector with the biggest eigenvalue obtained when analyzing all points in the
cluster, while the local orientations are obtained when considering specific subsets
of these points.

Figure 3.2 illustrates this fact for a toy 2-dimensional database containing one
cluster that follows a sinusoidal curve. As it can be seen, when considering specific
parts of the cluster, one can find considerably distinct local orientations, as in the
cases of the small neighborhoods centered at points a and b. The global orientation
is uncovered only when larger parts of the cluster are analyzed, as in the case of the
large neighborhood centered at the point c.

Therefore, pairs of microclusters that belong to the same correlation cluster and
have very distinct orientations are likely to exist, despite of the fact that they are
close to each other in one subspace. It is also possible that two microclusters of a
single cluster have similar orientations, although they are far apart in all subspaces.
Thus, the specification of levels of importance for the analyses of proximity and of
orientation in the merging process is a complex task and, at the same time, it is vital
to the definition of what microclusters should be merged to generate an accurate
clustering result. CURLER tackles this difficult problem by defining one similarity
measure for pairs of microclusters, named *co-sharing level*.

The algorithm searches for microclusters based on the Fuzzy Logic and on the
Expectation/Maximization (EM) technique [11]. Similarly to the k-means method,
the strategy proposed finds k_0 clusters (in this case named microclusters) through
an iterative process that aims at maximizing the quality of the microclusters found.

The value of k_0 is user-defined and the microclusters found are not necessarily disjoint. Quality is measured based on the adequacy of the microclusters found with regard to a predefined clustering model. The clustering model used is a Gaussian mixture model in which each microcluster M_i is represented by a probability distribution with density parameters $\theta_i = \{\mu_i, \sum_i\}$, where μ_i and $\sum_i$ represent respectively the centroid and the covariance matrix of the data elements that belong to M_i. A vector W is also defined, where each value W_i represents the fraction of the database that belongs to each microcluster M_i.

The iterative process begins with the creation of k_0 microclusters using identity matrices and random vectors to represent their covariance matrices and centroids respectively. The vector W is initially built by setting $W_i = 1/k_0$ for each microcluster M_i created. Similarly to any method based on the Expectation/Maximization (EM) technique, the iterative process has two steps per iteration: the *Expectation step* and the *Maximization step*. The first one computes the probabilities of each data element to belong to each microcluster. In the second step, the density parameters of each microcluster are updated. The process stops when, after one iteration, the increase in the clustering quality is smaller than the parameter ε*likelihood* or the maximum number of iterations *MaxLoopNum* is reached. The values for ε*likelihood* and *MaxLoopNum* are user-defined. The clustering quality obtained after one iteration is computed by $E\left(\theta_1, \theta_2, \ldots \theta_{k_0}|D\right) = \sum_{x\in D} log\left[\sum_{i=1}^{k_0} W_i.PR\left(M_i|x\right)\right]$, where x is an element of the database D and $PR\left(M_i|x\right)$ is the probability of an element x to belong to microcluster M_i. The final result of the iterative process is a set of k_0 microclusters, defined by their corresponding centroids and covariance matrices, as well as the probabilities of each data element to belong to each microcluster.

Once the iterative process is complete, the resulting microclusters are analyzed to define the ones that should be merged. For this task, CURLER proposes a similarity measure for pairs of microclusters, named *co-sharing level*. It is computed by the function $coshare(M_i, M_j) = \sum_{x\in D}\left[PR\left(M_i|x\right).PR\left(M_j|x\right)\right]$, where x is an element of the database D, $PR\left(M_i|x\right)$ is the probability of element x to belong to microcluster M_i and $PR\left(M_j|x\right)$ is the probability of element x to belong to the microcluster M_j. Based on the *coshare*( ) function, a recursive process merges pairs of microclusters to form clusters, including merged clusters in further mergers. Specifically, a pair of microclusters (M_i, M_j) is merged if and only if $coshare(M_i, M_j) \geq$ ε*coshare*, where the value of ε*coshare* is user-defined.

After that, the resulting clusters are refined in one semi-automatic process that involves their visualization. This step is motivated by the fact that the automatic process may tend to spot only the global orientations of the clusters and the visualization step allows the user to find clusters that need to be further explored. We shall omit the details of this process here, as this book focuses on automatic knowledge discovery methods only.

The main shortcomings of CURLER are: (1) the process is not fully automated; (2) the method disregards the possible existence of outliers in the data, which is considered to be a drawback, since existing methods for outlier detection do not analyze subspaces of the original space; (3) there is no guarantee that the iterative process will converge in feasible time for all cases. Notice that the user-defined maximum number of iterations does not actually solves the problem; and (4) the automatic part of the process has one quadratic time complexity regarding the number of microclusters and their dimensionalities, while the time complexity is cubic with respect to the data dimensionality.

3.5 Conclusions

In this chapter we analyzed some of the well-known methods from literature that spot clusters in multi-dimensional data with more than five or so dimensions. This analysis aimed at identifying the strategies already used to tackle the problem, as well as the major limitations of the existing, state-of-the-art techniques. Several well-known methods were briefly commented and concise descriptions of three of the most relevant ones were presented. One detailed survey on this area is in [23].

We conclude the chapter by summarizing in Table 3.1 the qualities of some of the most relevant techniques with regard to the main desirable properties that any clustering technique for moderate-to-high dimensionality data should have. Notice that Table 3.1 was inspired[1] in one table found in [23]. Specifically, Table 3.1 uses checkmarks to link techniques with their desirable properties in order to help one to evaluate and to compare the techniques listed. In general, the more checkmarks an algorithm features, the better and/or the more general is the algorithm. The properties considered are described as follows:

- **Arbitrarily oriented clusters**: the ability to spot clusters that exist in subspaces formed by linear combinations of the original dimensions, besides the ones existing in subspaces that refer to subsets of the original dimensions;
- **Not relying on the** *locality assumption*: the ability to be independent of the *locality assumption*, i.e., not to assume that the analysis of the space with all axes is sufficient to spot patterns that lead to clusters that only exist in subspaces;
- **Adaptive density threshold**: the ability to be independent of a fixed density threshold, i.e., not to assume high dimensional clusters to be as dense as low dimensional ones;
- **Independent of the order of the attributes**: the ability to generate the exact same clustering result for a given dataset, regardless of the order of its attributes;
- **Independent of the order of the objects**: the ability to generate the exact same clustering result for a given dataset, regardless of the order of its objects;
- **Deterministic**: the ability to generate the exact same clustering result for a given dataset every time that the algorithm is executed;

---

[1] Table 3.1 includes a summary of one table found in [23], i.e., Table 3.1 includes a selection of most relevant desirable properties and most closely related works from the original table. Table 3.1 also includes two novel desirable properties not found in [23]–**Linear or quasi-linear complexity** and **Terabyte-scale data analysis**.

Table 3.1 Properties of clustering algorithms well-suited to analyze moderate-to-high dimensional data

Clustering Algorithm	Arbitrarily oriented clusters	Not relying on the locality assumption	Adaptive density threshold	Independent of the order of the attributes	Independent of the order of the objects	Deterministic	Arbitrary number of clusters	Overlapping clusters (soft clustering)	Arbitrary subspace dimensionality	Avoiding complete enumeration	Robust to noise	Linear or quasi-linear complexity	Terabyte-scale data analysis
Axes parallel clustering													
CLIQUE [7, 8]		✓		✓	✓	✓	✓	✓	✓	✓		✓	
ENCLUS [16]		✓		✓	✓	✓	✓	✓	✓			✓	
SUBCLU [24]		✓		✓	✓	✓	✓	✓	✓			✓	
PROCLUS [5]			✓							✓			
PreDeCon [14]				✓	✓	✓	✓				✓	✓	
P3C [26, 27]	✓	✓		✓	✓	✓	✓		✓	✓		✓	
COSA [20]				✓	✓	✓					✓	✓	✓
DOC/ FASTDOC [29]	✓			✓	✓		✓	✓	✓				
FIRES [22]	✓	✓		✓	✓	✓	✓	✓	✓		✓	✓	
Correlation clustering													
ORCLUS [6, 4]	✓			✓						✓			
4C [15]	✓			✓	✓	✓	✓				✓	✓	
COPAC [2]	✓			✓	✓	✓	✓			✓	✓	✓	
CASH [1]	✓	✓	*n a*		✓	✓	✓			✓		✓	

Notice that this table was inspired in one table found in [23]. *n a*: not applicable

- **Arbitrary number of clusters**: the ability to automatically identify the number of clusters existing in the input dataset, i.e., not to take the number of clusters as an user-defined input parameter;
- **Overlapping clusters (soft clustering)**: the ability to identify data objects that belong to two or more overlapping clusters;
- **Arbitrary subspace dimensionality**: the ability to automatically spot clusters in subspaces of distinct dimensionalities in a single run of the algorithm, without taking the dimensionality of the clusters as an user-defined input parameter;
- **Avoiding complete enumeration**: the ability to avoid the analysis of all possible subspaces of a d-dimensional space, even for the worst case scenario;
- **Robust to noise**: the ability to obtain accurate clustering results from noisy data;

- **Linear or quasi-linear complexity**: the ability to scale linearly or quasi-linearly in terms of memory requirement and of execution time with regard to increasing numbers of points and axes, besides increasing clusters' dimensionalities;
- **Terabyte-scale data analysis**: the ability to handle datasets of Terabyte-scale in feasible time.

In spite of the several qualities of existing works, the analysis of the literature summarized in Table 3.1 leads us to identify one important issue. To the best of our knowledge, among the methods published in the literature and designed to look for clusters in subspaces, no one has *any* of the following desirable properties:

- **Linear or quasi-linear complexity**
- **Terabyte-scale data analysis**

A complete justification for this statement is found in the descriptions provided during this chapter and also in the previous Chap. 2. Especially, please note that the existing techniques are unable to cluster datasets of Terabyte-scale in feasible time mainly because they propose serial processing strategies, besides the fact that they do not provide linear or quasi-linear scalability. Details on this topic are found in Sect. 2.5.

The next three chapters contain the central part of this book. They present detailed descriptions of three novel knowledge discovery methods aimed at tackling the problem of mining large sets of complex data in the scale of Terabytes.

References

1. Achtert, E., Böhm, C., David, J., Kröger, P., Zimek, A.: Global correlation clustering based on the hough transform. Stat. Anal. Data Min. **1**, 111–127 (2008)
2. Achtert, E., Böhm, C., Kriegel, H.P., Kröger, P., Zimek, A.: Robust, complete, and efficient correlation clustering. SDM, USA (2007)
3. Agarwal, P.K., Mustafa, N.H.: k-means projective clustering. In: PODS, pp. 155–165. ACM, Paris, France (2004). http://doi.acm.org/10.1145/1055558.1055581
4. Aggarwal, C., Yu, P.: Redefining clustering for high-dimensional applications. IEEE TKDE **14**(2), 210–225 (2002). http://doi.ieeecomputersociety.org/10.1109/69.991713
5. Aggarwal, C.C., Wolf, J.L., Yu, P.S., Procopiuc, C., Park, J.S.: Fast algorithms for projected clustering. SIGMOD Rec. **28**(2), 61–72 (1999). http://doi.acm.org/10.1145/304181.304188
6. Aggarwal, C.C., Yu, P.S.: Finding generalized projected clusters in high dimensional spaces. SIGMOD Rec. **29**(2), 70–81 (2000). http://doi.acm.org/10.1145/335191.335383
7. Agrawal, R., Gehrke, J., Gunopulos, D., Raghavan, P.: Automatic subspace clustering of high dimensional data for data mining applications. SIGMOD Rec. **27**(2), 94–105 (1998). http://doi.acm.org/10.1145/276305.276314
8. Agrawal, R., Gehrke, J., Gunopulos, D., Raghavan, P.: Automatic subspace clustering of high dimensional data. Data Min. Knowl. Discov. **11**(1), 5–33 (2005). doi:10.1007/s10618-005-1396-1
9. Aho, A.V., Hopcroft, J.E., Ullman, J.: The Design and Analysis of Computer Algorithms. Addison-Wesley Longman Publishing Co., Inc., Boston (1974)
10. Al-Razgan, M., Domeniconi, C.: Weighted clustering ensembles. In: J. Ghosh, D. Lambert, D.B. Skillicorn, J. Srivastava (eds.) SDM. SIAM (2006)

11. Banfield, J.D., Raftery, A.E.: Model-based gaussian and non-gaussian clustering. Biometrics **49**(3), 803–821 (1993)
12. Böhm, C., Faloutsos, C., Pan, J.Y., Plant, C.: Robust information-theoretic clustering. In: KDD, pp. 65–75. USA (2006). http://doi.acm.org/10.1145/1150402.1150414
13. Böhm, C., Faloutsos, C., Plant, C.: Outlier-robust clustering using independent components. In: SIGMOD, pp. 185–198. USA (2008). http://doi.acm.org/10.1145/1376616.1376638
14. Bohm, C., Kailing, K., Kriegel, H.P., Kroger, P.: Density connected clustering with local subspace preferences. In: ICDM '04: Proceedings of the Fourth IEEE International Conference on Data Mining, pp. 27–34. IEEE Computer Society, Washington (2004)
15. Böhm, C., Kailing, K., Kröger, P., Zimek, A.: Computing clusters of correlation connected objects. In: SIGMOD, pp. 455–466. USA (2004). http://doi.acm.org/10.1145/1007568.1007620
16. Cheng, C.H., Fu, A.W., Zhang, Y.: Entropy-based subspace clustering for mining numerical data. In: KDD, pp. 84–93. USA (1999). http://doi.acm.org/10.1145/312129.312199
17. Cheng, H., Hua, K.A., Vu, K.: Constrained locally weighted clustering. In: Proceedings of the VLDB **1**(1), 90–101 (2008). http://doi.acm.org/10.1145/1453856.1453871
18. Domeniconi, C., Gunopulos, D., Ma, S., Yan, B., Al-Razgan, M., Papadopoulos, D.: Locally adaptive metrics for clustering high dimensional data. Data Min. Knowl. Discov. **14**(1), 63–97 (2007). doi:10.1007/s10618-006-0060-8
19. Domeniconi, C., Papadopoulos, D., Gunopulos, D., Ma, S.: Subspace clustering of high dimensional data. In: M.W. Berry, U. Dayal, C. Kamath, D.B. Skillicorn (eds.) SDM (2004)
20. Friedman, J.H., Meulman, J.J.: Clustering objects on subsets of attributes (with discussion). J. R. Stat. Soc. Ser. B **66**(4), 815–849 (2004). doi:ideas.repec.org/a/bla/jorssb/v66y2004i4p815-849.html
21. Grunwald, P.D., Myung, I.J., Pitt, M.A.: Advances in Minimum Description Length: Theory and Applications (Neural Information Processing). The MIT Press, Cambridge (2005)
22. Kriegel, H.P., Kröger, P., Renz, M., Wurst, S.: A generic framework for efficient subspace clustering of high-dimensional data. In: ICDM, pp. 250–257. Washington (2005). http://dx.doi.org/10.1109/ICDM.2005.5
23. Kriegel, H.P., Kröger, P., Zimek, A.: Clustering high-dimensional data: A survey on subspace clustering, pattern-based clustering, and correlation clustering. ACM TKDD **3**(1), 1–58 (2009). doi:10.1145/1497577.1497578
24. Kröger, P., Kriegel, H.P., Kailing, K.: Density-connected subspace clustering for high-dimensional data. In: SDM, USA (2004)
25. Moise, G., Sander, J.: Finding non-redundant, statistically significant regions in high dimensional data: a novel approach to projected and subspace clustering. In: KDD, pp. 533–541 (2008)
26. Moise, G., Sander, J., Ester, M.: P3C: A robust projected clustering algorithm. In: ICDM, pp. 414–425. IEEE Computer Society (2006)
27. Moise, G., Sander, J., Ester, M.: Robust projected clustering. Knowl. Inf. Syst. **14**(3), 273–298 (2008). doi:10.1007/s10115-007-0090-6
28. Ng, E.K.K., chee Fu, A.W., Wong, R.C.W.: Projective clustering by histograms. TKDE **17**(3), 369–383 (2005). doi:10.1109/TKDE.2005.47
29. Procopiuc, C.M., Jones, M., Agarwal, P.K., Murali, T.M.: A monte carlo algorithm for fast projective clustering. In: SIGMOD, pp. 418–427. USA (2002). http://doi.acm.org/10.1145/564691.564739
30. Rissanen, J.: Stochastic Complexity in Statistical Inquiry Theory. World Scientific Publishing Co., Inc., River Edge (1989)
31. Tung, A.K.H., Xu, X., Ooi, B.C.: Curler: finding and visualizing nonlinear correlation clusters. In: SIGMOD, pp. 467–478 (2005). http://doi.acm.org/10.1145/1066157.1066211
32. Yip, K., Cheung, D., Ng, M.: Harp: a practical projected clustering algorithm. TKDE **16**(11), 1387–1397 (2004)
33. Yiu, M.L., Mamoulis, N.: Iterative projected clustering by subspace mining. TKDE **17**(2), 176–189 (2005)

Chapter 4
Halite

Abstract In the previous chapter, we provide one concise description of some of the representative methods for clustering moderate-to-high-dimensional data, and summarize the analysis of the literature in Table 3.1. It allows us to identify two main desirable properties that are still missing from the existing techniques: (1) **Linear or quasi-linear complexity** and; (2) **Terabyte-scale data analysis**. Here we describe one work that focuses on tackling the former problem. Specifically, this chapter presents the new **method** *Halite* **for correlation clustering** [4, 5]. *Halite* is a novel *correlation clustering* method for multi-dimensional data, whose main strengths are that it is fast and scalable with regard to increasing numbers of objects and axes, besides increasing dimensionalities of the clusters. The following sections describe the new method in detail.

Keywords Correlation clustering · Hard clustering · Soft clustering · Complex data · Linear or quasi-linear complexity · Support for breast cancer diagnosis · Satellite imagery analysis

4.1 Introduction

The **method** *Halite* **for correlation clustering** [4, 5] is a fast and scalable algorithm that looks for *correlation clusters* in multi-dimensional data using a *top-down*, multi-resolution strategy. It analyzes the distribution of points in the space with all dimensions by performing a multi-resolution, recursive partitioning of that space, which helps distinguishing clusters covering regions with varying sizes, density, correlated axes and number of points. Existing methods are typically super-linear in either space or execution time. *Halite* is fast and scalable, and gives highly accurate results. In details, the main contributions of *Halite* are:

1. **Scalability**: it is linear in running time and in memory usage with regard to the data size and to the dimensionality of the subspaces where clusters exist. *Halite* is

R. L. F. Cordeiro et al., *Data Mining in Large Sets of Complex Data*,
SpringerBriefs in Computer Science, DOI: 10.1007/978-1-4471-4890-6_4,
© The Author(s) 2013

Fig. 4.1 Example of an isometric crystal system commonly found in the nature—one specimen of the mineral halite that was naturally crystallized. Notice that this specimen was obtained from the stassfurt potash deposit, Saxony–Anhalt, Germany. Image courtesy of Rob Lavinsky, iRocks.com—CC-BY-SA-3.0

also linear in memory usage and quasi-linear in running time regarding the space dimensionality;

2. **Usability**: it is deterministic, robust to noise, does not have the number of clusters as a parameter and finds clusters in subspaces generated by the original axes or by their linear combinations, including space rotation;

3. **Effectiveness**: it is accurate, providing results with equal or better quality compared to top related works;

4. **Generality**: it includes a soft clustering approach, which allows points to be part of two or more clusters that overlap in the data space.

The new clustering method is named after the mineral Halite. Halite, or rock salt, is the mineral form of sodium chloride ($NaCl$). One specimen of naturally crystallized Halite is shown in Fig. 4.1. Halite forms isometric crystals, i.e., crystal systems of any shape and size formed from the union of overlapping, rectangular unit cells. The new clustering method proposes to generalize the structure of these systems to the d-dimensional case in order to describe correlation clusters of any shape and size, hence its name *Halite*.

The method *Halite* uses spatial convolution masks in a novel way to efficiently detect density variations in a multi-scale grid structure that represents the input data, thus spotting clusters. These masks are extensively used in digital image processing to detect patterns in images [7]. However, to the best of our knowledge, this is the first work to apply such masks over data in five or more axes. *Halite* also uses the Minimum Description Length (MDL) principle [8, 14] in a novel way. The main idea is to encode an input dataset, selecting a minimal code length. Specifically, *Halite* uses MDL to automatically tune a density threshold with regard to the data distribution, which helps spotting the clusters' subspaces. Finally, *Halite* includes a compression-based analysis to spot points that most likely belong to two or more clusters that overlap in the space. It allows soft clustering results, i.e., points can be part of two or more overlapping clusters.

One theoretical study on the time and space complexity of *Halite*, as well as an extensive experimental evaluation performed over synthetic and real data spanning up to 1 *million* elements corroborate the desirable properties of the new method regarding its Scalability, Usability, Effectiveness and Generality. Specifically, experiments comparing *Halite* with seven representative works were performed. On synthetic data, *Halite* was consistently the fastest method, always presenting highly accurate results. Regarding real data, *Halite* analyzed 25-dimensional data for breast cancer diagnosis (KDD Cup 2008) at least 11 times faster than five previous works, increasing their accuracy in up to 35 %, while the last two related works failed. Details are found in the upcoming Sect. 4.7.

Remark: *Halite* is well-suited to analyze data in the range of 5–30 axes. The intrinsic dimensionalities of real datasets are frequently smaller than 30, mainly due to the existence of several global correlations [15]. Therefore, if one dataset has more than 30 or so axes, it is possible to apply some distance preserving dimensionality reduction or feature selection algorithm to remove the global correlations, such as PCA [9] or FDR [16], and then apply *Halite* to treat the correlations local to specific data clusters.

4.2 General Proposal

Halite tackles the problem of correlation clustering based on the following fact: clusters of *any* shape, existing only in subspaces of a d-dimensional space, create bumps (spikes) in the point density of their respective subspaces, but, commonly, these bumps *still exist* in the space with all dimensions, besides being weakened or diluted by the dimensions that do not belong to the respective clusters' subspaces. To find the clusters, in the same way as most related works do (see Chap. 3 for details), *Halite* assumes what is known in literature as the *locality assumption*: one can still spot such diluted bumps by analyzing the space with all dimensions.

Assuming *locality*, and inspired by the structure of isometric crystal systems, such as the specimen of the mineral Halite shown in Fig. 4.1, the new method's general proposal for correlation clustering is twofold:

- **Bump Hunting:** to spot bumps in the point density of the space with all axes, defining each bump as a d-dimensional, axes-aligned hyper-rectangle that is

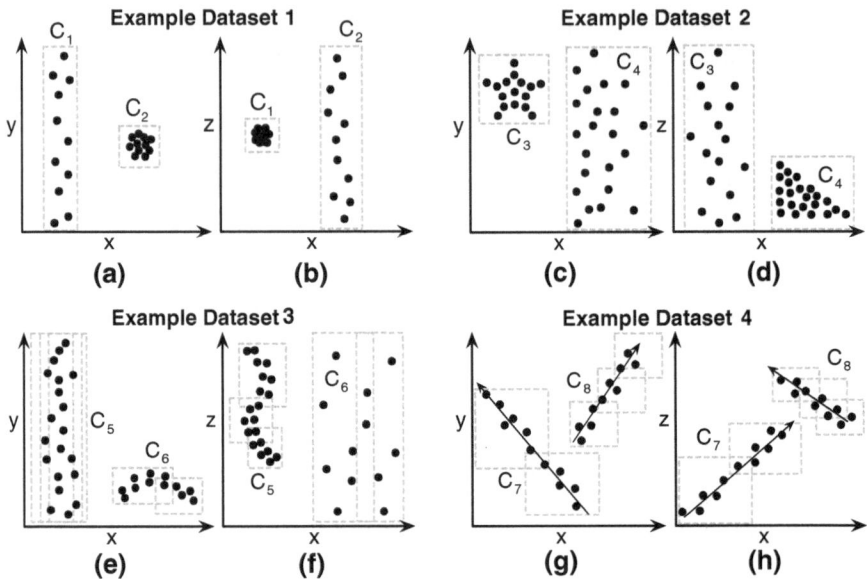

Fig. 4.2 $x - y$ and $x - z$ projections of four 3-dimensional datasets over axes $\{x, y, z\}$. From **a–f** clusters in the subspaces $\{x, y\}$ and $\{x, z\}$. **g** and **h** clusters in subspaces formed by linear combinations of $\{x, y, z\}$

considerably denser than its neighboring space regions (local neighborhood). Each bump is named as a β-cluster;

- **Correlation Clustering:** to describe correlation clusters of *any* shape, existing in subspaces of a d-dimensional space, axis-aligned or not, as maximal sets of overlapping β-clusters.

Let us illustrate these concepts on the 3-dimensional toy datasets that were previously introduced in Sect. 2.3. Figure 4.2 reprints the same datasets in this chapter, this time *including* dotted rectangles into the illustration to represent β-clusters. Remember: the datasets contain clusters of several shapes (i.e., elliptical clusters, 'star-shaped' clusters, 'triangle-shaped' clusters and 'curved' clusters) existing only in subspaces that are formed by subsets of the axes $\{x, y, z\}$, or by linear combinations of these axes, and we plot x-y and x-z projections of the toy data in the illustration. Notice that a single axes-aligned, 3-dimensional hyper-rectangle embodies each cluster in some cases, but certain clusters, especially the ones not aligned to the original axes, need more than one overlapping hyper-rectangle to be covered. Thus, maximal sets of overlapping β-clusters properly describe the clusters, even the ones not aligned to the axes $\{x, y, z\}$.

According to the general proposal presented, formal definitions for the β-clusters and for the correlation clusters are provided.

Definition 4.1 Let $^d S$ be a multi-dimensional dataset on the axes E. Then a β-**cluster** in $^d S$, $^\delta_\beta C_k = \langle _\beta E_k, {}^\delta_\beta S_k \rangle$ is a set $_\beta E_k \subseteq E$ of δ axes together with a set of points $^\delta_\beta S_k \subseteq {}^d S$ that, in a statistical sense, form one **bump** (one spike) in the point density of the d-dimensional space. The axes in $_\beta E_k$, **relevant to the** β-**cluster**, are the original axes that cause the bump, i.e., $^\delta_\beta S_k$ is densely clustered in the axes-aligned subspace formed by $_\beta E_k$. The axes in $E -_\beta E_k$ are **irrelevant to** $^\delta_\beta C_k$. $^\delta_\beta S_k$ must be within an axes-aligned, d-dimensional hyper-rectangle, with the upper and lower bounds at each axis e_j being $U[k][j]$ and $L[k][j]$ respectively.

Definition 4.2 β-**cluster overlapping:** Given any two β-clusters $^{\delta'}_\beta C_{k'}$ and $^{\delta''}_\beta C_{k''}$, one can say that the β-clusters overlap to each other if $U[k'][j] \geq L[k''][j] \wedge L[k'][j] \leq U[k''][j]$ is valid for every original axis e_j.

Definition 4.3 Let $^d S$ be a multi-dimensional dataset on the axes E. Then a **correlation cluster** in $^d S$, $^\delta_\gamma C_k = \langle _\gamma E_k, {}^\delta_\gamma S_k \rangle$ is defined as one maximally connected component in the graph, whose nodes are the β-clusters that exist in $^d S$, and there is an edge between two nodes, if the respective β-clusters overlap. For the β-clusters referring to the nodes of this component, $^\delta_\gamma S_k = \bigcup {}^{\delta'}_\beta S_{k'}$ and $_\gamma E_k = \bigcup {}_\beta E_{k'}$. The axes in $_\gamma E_k$ are said to be **relevant to the cluster**, and the axes in $E -_\gamma E_k$ are **irrelevant to the cluster**. The cardinality $\delta = |_\gamma E_k|$ is the **dimensionality of the cluster**.

Notice one *important* remark: as it can be seen in Definition 4.3, *Halite* assumes that the axes irrelevant to a cluster $^\delta_\gamma C_k = \langle _\gamma E_k, {}^\delta_\gamma S_k \rangle$ are the original axes in which $^\delta_\gamma C_k$ spreads parallel to. The axes relevant to $^\delta_\gamma C_k$ are the remaining *original* axes. Thus, if $^\delta_\gamma C_k$ does *not* spread parallel to any original axis, *all* original axes are relevant to the cluster, i.e., $_\gamma E_k = E$. Finally, once a cluster $^\delta_\gamma C_k$ is found, and thus its corresponding local correlation was already detected, the **subspace** in which the cluster exists can be easily defined by using a traditional feature extraction method, such as PCA, applied over the points $^\delta_\gamma S_k$ projected onto the axes $_\gamma E_k$. $_\gamma E_k$ defines the subspace of a cluster only for axes-aligned clusters. As an example, cluster C_6 in Fig. 4.2 spreads parallel to z, while both its relevant axes and subspace are given by the axes $\{x, y\}$. On the other hand, cluster C_8 in Fig. 4.2 does not spread parallel to any original axis, thus its relevant axes are $\{x, y, z\}$. The subspace of C_8 is found by feature extraction applied only on the points of the cluster, projected into its relevant axes $\{x, y, z\}$.

4.3 Presented Method: Basics

This section presents the basic implementation of the new clustering method,[1] which is referred to as *Halite*$_0$. We start with *Halite*$_0$ for clarity of description; in later

---

[1] One first implementation of *Halite*$_0$ was initially named as the method *MrCC* (after Multi-resolution Correlation Clustering) in [4]. Latter, it was renamed to *Halite*$_0$ for clarity, since several improvements on the basic implementation were included into one journal paper [5].

sections, we describe the additional improvements, that lead to the optimized, and finally recommended method *Halite*.

The main idea in the new approach is to identify clusters based on the variation of the data density over the space in a multi-resolution way, dynamically changing the partitioning size of the analyzed regions. Multi-resolution is explored applying d-dimensional hyper-grids with cells of several side sizes over the data space and counting the points in each grid cell. Theoretically, the number of cells increases exponentially to the dimensionality as the cell size shrinks, so the grid sizes dividing each region are carefully chosen in a way that only the cells that hold at least one data element are stored, limiting this number to the dataset cardinality. The grid densities are stored in a quad-tree-like data structure, the Counting-tree, where each level represents the data as a hyper-grid in a specific resolution.

Spatial convolution masks are applied over each level of the Counting-tree, to identify bumps in the data distribution regarding each resolution. Applying the masks to the needed tree levels allows spotting clusters with different sizes. Given a tree level, $Halite_0$ applies a mask to find the regions of the space with all dimensions that refer to the largest changes in the point density. The regions found may indicate clusters that only exist in subspaces of the analyzed data space. The neighborhoods of these regions are analyzed to define if they stand out in the data in a statistical sense, thus indicating clusters. The axes in which the points in an analyzed neighborhood are close to each other are said to be relevant to the respective cluster, while the axes in which these points are sparsely distributed are said to be irrelevant to the cluster. The Minimum Description Length (MDL) principle is used in this process to automatically tune a threshold able to define relevant and irrelevant axes, based on the data distribution. The following subsections detail the method.

4.3.1 Building the Counting-Tree

The first phase of $Halite_0$ builds the Counting-tree, representing a dataset $^d S$ with d axes and η points as a set of hyper-grids of d-dimensional cells in several resolutions. The tree root (level zero) corresponds to a hyper-cube embodying the full dataset. The next level divides the space into a set of 2^d hyper-cubes, each of which fulfilling a "hyper-quadrant" whose side size is equal to half the size of the previous level. The resulting hyper-cubes are divided again, generating the tree structure. Therefore, each level h of the Counting-tree represents $^d S$ as a hyper-grid of d-dimensional cells of side size $\xi_h = 1/2^h$, $h = 0, 1, ..., H - 1$, where H is the number of resolutions. Each cell can either be refined in the next level or not, according to the presence or to the absence of points in the cell, so the tree can be unbalanced.

Without loss of generality, we assume in the following description that the dataset $^d S$ is in a unitary hyper-cube $[0, 1)^d$. Thus, we refer to each cell in the Counting-tree as a hyper-cube. However, for data not scaled between 0 and 1, as long as one knows the minimum and the maximum bounds of the axes, the Counting-tree can still be built by dividing each axis in half, through a recursive procedure that creates cells referring to hyper-rectangles and all the strategies described still apply.

The structure of each tree cell is defined as $< loc, n, P[d], usedCell, ptr >$, where loc is the cell spatial position inside its parent cell, n is the number of points in the cell, $P[\,]$ is a d-dimensional array of half-space counts, $usedCell$ is a boolean flag and ptr is a pointer to the next tree level. The cell position loc locates the cell inside its parent cell. It is a binary number with d bits of the form $[bb \ldots b]$, where the j-bit sets the cell in the lower (0) or upper (1) half of axis e_j relative to its immediate parent cell. Each half-space count $P[j]$ counts the number of points in the lower half of the cell with regard to axis e_j. The flag $usedCell$ determines whether or not the cell's density of points has already been analyzed in the clustering procedure. This analysis occurs only in the second phase of $Halite_0$, and thus, every $usedCell$ flag is initially set to *false*. Figure 4.3a shows 2-dimensional grids in up to five resolutions, while Fig. 4.3b and c respectively illustrate a grid over a 2-dimensional dataset in four resolutions and the respective Counting-tree. The $usedCell$ flags are not shown to reduce cluttering in the figure.

In a Counting-tree, a given cell a at level h is referred to as a_h. The immediate parent of a_h is a_{h-1} and so on. The cell position loc of a_h corresponds to the relative position regarding its immediate parent cell, and it is referred to as $a_h.loc$. The parent cell is at relative position $a_{h-1}.loc$ and so on. Hence, the absolute position of cell a_h is obtained by following the sequence of positions $\left[a_1.loc \downarrow a_2.loc \downarrow \ldots \downarrow a_{h-1}.loc \downarrow a_h.loc \right]$. For example, the cell marked as A_2 in level 2 of Fig. 4.3c has relative position $A_2.loc = [00]$ and absolute position up to level 2 as $[11 \downarrow 00]$. A similar notation is used to refer to the other cells' attributes, n, $P[\,]$, $usedCell$ and ptr. As an example, given the cell a_h, its number of points is $a_h.n$, the number of points in its parent cell is $a_{h-1}.n$ and so on.

The Counting-tree is created in main memory, and an efficient implementation to deal with the case when memory is not enough is described in Sect. 4.4. Each tree node is implemented as a linked list of cells. Therefore, although the number of regions to divide the space grows exponentially at $O(2^{dH})$, the clustering method only stores the regions where there is at least one point and each tree level has in fact at most η cells. However, without loss of generality, this section describes each node as an array of cells for clarity.

Algorithm 1 shows how to build a Counting-tree. It receives the dataset normalized to a d-dimensional hyper-cube $[0, 1)^d$ and the desired number of resolutions H. This number must be greater than or equal to 3, as the tree contains $H - 1$ levels and at least 2 levels are required to look for clusters (see details in Algorithms 1 and 2). $Halite_0$ performs a single data scan, counting each point in every corresponding cell at each tree level as it is read. Each point is also counted in each half-space count $P[j]$ for every axis e_j when it is at the lower half of a cell in e_j.

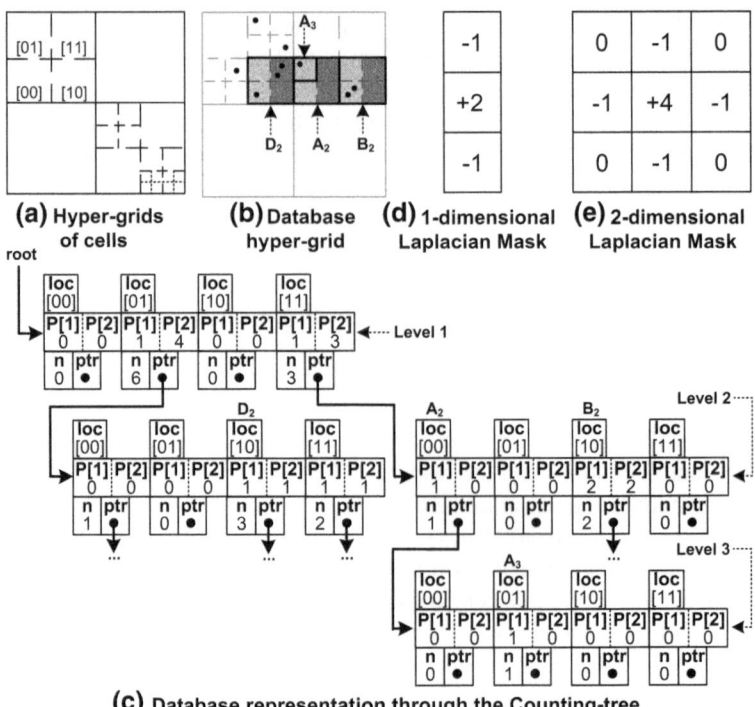

Fig. 4.3 Examples of Laplacian masks, 2-dimensional hyper-grid cells and the corresponding counting-tree. Grids of several resolutions are applied over the dataset in (**b**) and the tree in (**c**) stores the resulting point counts. A convolution mask helps spotting correlation clusters

---

Algorithm 1 : Building the Counting-tree.

Input: normalized dataset ^dS,
         number of distinct resolutions H
Output: Counting-tree T
1: **for** each point $s_i \in {}^dS$ **do**
2:    start at the root node;
3:    **for** $h = 1, 2, ..., H - 1$ **do**
4:       decide which hyper-grid cell in the current node of the Counting-tree covers s_i (let it be the cell a_h);
5:       $a_h.n = a_h.n + 1$;
6:       $a_h.usedCell = false$;
7:       if $h > 1$, update the half-space counts in a_{h-1};
8:       access the tree node pointed by $a_h.ptr$;
9:    **end for**
10:   update the half-space counts in a_h;
11: **end for**

4.3.1.1 Time and Space Complexity

Algorithm 1 reads each of the η data points once. When a point is read, it is counted in all the $H - 1$ tree levels, based on its position in all the d axes. Thus, the time complexity of Algorithm 1 is $O(\eta\ H\ d)$. The tree has $H - 1$ levels. Each level has at most η cells, which contain an array with d positions each. Thus, the space complexity of Algorithm 1 is $O(H\ \eta\ d)$.

4.3.2 Finding β-Clusters

The second phase of *Halite$_0$* uses the counts in the tree to spot bumps in the space with all axes that indicate β-clusters. A β-cluster $\,_\beta^\delta C_k = \left\langle \,_\beta E_k, \,_\beta^\delta S_k \right\rangle$ follows Definition 4.1. *Halite$_0$* uses three matrices L, U and V to describe β-clusters. Let $\,_\beta k$ be the number of β-clusters found so far. Each matrix has $\,_\beta k$ lines and d columns, and the description of a β-cluster $\,_\beta^\delta C_k$ is in arrays $L[k]$, $U[k]$ and $V[k]$. $L[k]$ and $U[k]$ respectively store the lower and the upper bounds of β-cluster $\,_\beta^\delta C_k$ in each axis, while $V[k]$ has the value *true* in $V[k][j]$ if axis e_j belongs to $\,_\beta E_k$, and the value *false* otherwise.

Halite$_0$ looks for β-clusters by applying convolution masks over each level of the Counting-tree. This task is named as *"Bump Hunting"*. The masks are integer approximations of the Laplacian filter, a second-derivative operator that reacts to transitions in density. Figures 4.3d and e show examples of 1- and 2-dimensional Laplacian masks respectively. In a nutshell, the *"Bump Hunting"* task refers to: apply for each level of the Counting-tree one d-dimensional Laplacian mask over the respective grid to spot bumps in the respective resolution. Figure 4.4a illustrates the process on a toy 1-dimensional dataset with grids in five resolutions. To spot bumps, for each cell of each resolution, the 1-dimensional mask from Fig. 4.3d is applied as follows: multiply the count of points of the cell by the center value in the mask; multiply the point count of each neighbor of the cell by the respective mask value, and; get the convoluted value for the cell by summing the results of the multiplications. After visiting all cells in one resolution, the cell with the largest convoluted value represents the clearest bump in that resolution, i.e., the largest positive magnitude of the density gradient. In Fig. 4.4a, for each resolution, one dark-gray arrow points to this cell.

For performance purposes, *Halite$_0$* uses only masks of order 3, that is, matrices of sizes 3^d. In such masks, regardless of their dimensionality, there is always one center element (the convolution pivot), $2d$ center-faces elements (or just face elements, for short) and $3^d - 2d - 1$ corner elements. Applying a mask over all cells at a Counting-tree level can become prohibitively expensive in datasets with several dimensions—for example, a 10-dimensional cell has 59,028 corner elements. However, *Halite$_0$* uses Laplacian masks having non-zero values only at the center and the facing elements, that is $2d$ for the center and -1 for the face elements, as in the

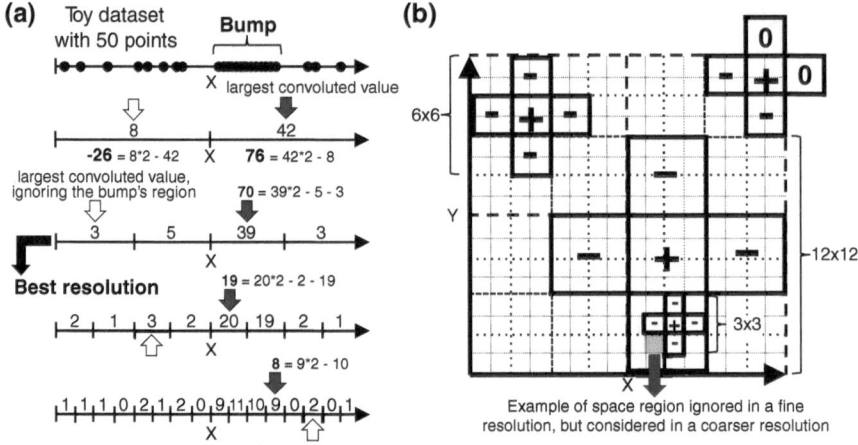

Fig. 4.4 **a** one mask applied to 1-dimensional data. An statistical test finds the best resolution for a bump, and the time to stop spotting bumps, **b** intuition on the success of the used masks. The multi-resolution allows using 3^d masks to "simulate" larger masks, while the space regions ignored by the mask borders in a resolution tend to be considered in coarser resolutions

examples in Fig. 4.3d and e. A 10-dimensional cell has only 20 face elements. Therefore, it is possible to convolute each level of a Counting-tree with linear complexity regarding the dataset dimensionality d.

Notice one *important* remark: *Halite*$_0$ uses masks of order 3, since they are the smallest available for convolution. Experimentally it was found that the clustering accuracy of *Halite*$_0$ improves a little (at most 5 % when applied over the datasets from the upcoming Sect. 4.7) when one uses masks of order ϕ, $\phi > 3$, having non-zero values at all elements (center, face and corner elements), but the time required increases too much—in the order of $O(\phi^d)$ as compared to $O(d)$ when using masks of order 3 having non-zero values only at the center and the facing elements. Thus, *Halite*$_0$ always uses masks of order $\phi = 3$. To explain the success of the used masks, we point to a fact: the multi-resolution allows *Halite*$_0$ to use masks of a low order to efficiently "simulate" masks of higher orders. Figure 4.4b gives an intuition on this fact by illustrating masks applied to grids in distinct resolutions over a 2-dimensional space. As it can be seen, one mask of order 3 applied to a coarse resolution "simulates" masks of higher orders (i.e., 6, 12, ...) applied to finer resolutions. Furthermore, Fig. 4.4b also shows that, in the multi-resolution setting used, the space regions ignored in fine resolutions due to the mask's zero values are commonly considered in coarser resolutions.

4.3.2.1 Bump Hunting

As it can be seen in Algorithm 2, phase two starts applying the mask to level two of the Counting-tree, starting at a coarse resolution and refining as needed. It allows

Halite$_0$ to find β-clusters of different sizes. When analyzing a resolution level h, the mask is convoluted over every cell at this level, excluding those already used for a β-cluster found before. The cell with the largest convoluted value refers to the clearest bump in the data space analyzed regarding level h, i.e., the largest positive magnitude of the density gradient. Thus, it may indicate a new β-cluster.

Applying a convolution mask to level h requires a partial walk over the Counting-tree, but no node deeper than h is visited. The walk starts going down from the root until reaching a cell b_h at level h that may need to be convoluted. The neighbors of this cell are named after the convolution matrix: face and corner neighbors. The center element corresponds to the cell itself. If the $b_h.usedCell$ flag is *true* or the cell shares the data space with a previously found β-cluster, the cell is skipped. Otherwise, *Halite$_0$* finds the face neighbors of b_h and applies the mask centered at b_h. After visiting all cells in level h, the cell with the largest convoluted value has the *usedCell* flag set to *true*.

Each cell b_h at resolution level h is itself a d-dimensional hyper-cube and it can be divided into other 2^d cells in the next level, splitting each axis in half. Therefore, from the two face neighbors at axis e_j, one is stored in the same node of cell b_h, while the other is in a sibling node of cell b_h. The face neighbor stored at the same node is named as the internal neighbor of cell b_h, and the other is named as its external neighbor regarding axis e_j. For example, cell A_2 in resolution level 2 of the Counting-tree in Fig. 4.3c has the internal neighbor B_2 and the external neighbor D_2 regarding axis e_1. The internal neighbor of a cell b_h in axis e_j at resolution level h and its point count are respectively referred as $NI(b_h, e_j)$ and $NI(b_h, e_j).n$. Similarly, for the same axis and resolution, the external neighbor of b_h and its points count are $NE(b_h, e_j)$ and $NE(b_h, e_j).n$ respectively. *Halite$_0$* analyzes the absolute cell positions in a Counting-tree to spot external and internal neighbors.

4.3.2.2 Confirming the β-Cluster

The *"Bump Hunting"* task allows one to efficiently spot the clearest bumps in a dataset. However, two open questions still prevent its use for clustering: (1) *what is the best resolution to spot each bump?* and (2) *when should the "Bump Hunting" stop?* To give an intuition on both questions, we use again the 1-dimensional data with grids in five resolutions from Fig. 4.4a. In the illustration, a dark-gray arrow points to the cell with the largest convoluted value, which describes the clearest bump in the data for each resolution. Notice: we did not describe yet a procedure to automatically identify the best resolution to spot a bump, which refers to question one. In the example, the bump is best described in the third coarsest resolution, as its bounds are overestimated in the two coarser resolutions and only its borders are spotted in the two finer resolutions. Now, let us refer to question two. Assuming that the exemplified bump was properly described in its best resolution, should *Halite$_0$* keep looking for bumps or should it stop? A procedure to automatically answer this question, after spotting each bump, is also missing. In the example, ignoring the bump's region, a white arrow points to the cell with the largest convoluted value

---

Algorithm 2 : Finding the β-clusters.

Input: Counting-tree T, significance level α
Output: matrices of β-clusters L, U and V, number of β-clusters $_\beta k$
1: $_\beta k = 0$;
2: **repeat**
3:    $h = 1$;
4:    **repeat**
5:       $h = h + 1$;
6:       **for** each cell b_h in level h of T **do**
7:          **if** $b_h.usedCell = false$ $\land$ b_h does not overlap with a previously found β-cluster **then**
8:             find the face neighbors of b_h in all axes;
9:             apply the Laplacian mask centered in b_h, using the point counts of b_h and of the found neighbors;
10:             $a_h = b_h$, if the resulting convoluted value is the largest one found in the current iteration;
11:          **end if**
12:       **end for**
13:       $a_h.usedCell = true$;
14:       centered on a_h and based on α, compute cP_j, nP_j and θ_j^α for every axis e_j;
15:       **if** $cP_j > \theta_j^\alpha$ for at least one axis e_j **then**
16:          $_\beta k = _\beta k + 1$; {a new β-cluster was found}
17:       **end if**
18:    **until** a new β-cluster is found $\lor$ $h = H - 1$
19:    **if** a new β-cluster was found **then**
20:       compute $r[\,]$ and $cThreshold$;
21:       **for** $j = 1, 2, ..., d$ **do**
22:          $V[_\beta k][j] = r[j] \geq cThreshold$;
23:          **if** $V[_\beta k][j] = true$ **then**
24:             compute $L[_\beta k][j]$ and $U[_\beta k][j]$ based on the lower and upper bounds of a_h and of its face neighbors regarding axis e_j;
25:          **else** $L[_\beta k][j] = 0$; $U[_\beta k][j] = 1$;
26:          **end if**
27:       **end for**
28:    **end if**
29: **until** no new β-cluster was found

---

for each resolution, which, clearly, does not lead to a cluster. Therefore, the *"Bump Hunting"* should stop.

Notice that the *"Bump Hunting"* spots cells on the "positive" side of the largest, local density changes, but it does not guarantee that these changes are statistically significant—some of them could have been created by chance. Even when analyzing only points randomly distributed through a d-dimensional space, the mask will return bumps. Therefore, *Halite$_0$* proposes to automatically answer both questions previously posed by ignoring bumps that, potentially, were created by chance, assuming that only statistically significant bumps lead to β-clusters. This strategy allows *Halite$_0$* to identify the best resolution to spot each bump, since this resolution is the one in which the corresponding local density change is more intense, thus, it avoids overestimating the clusters bounds and spotting only clusters' borders. The proposed strategy also spots the correct moment to stop the *"Bump Hunting"*—it stops once the clearest bump was potentially created by chance in all resolutions.

To state if a new β-cluster exists in level h, $Halite_0$ searches for the cell a_h with the largest convoluted value and analyzes its neighbors. The intuition is to use an statistical test to verify if the possible cluster, represented by cell a_h, is *significantly* denser than its neighboring data space regions, thus confirming or not the cluster existence. The Counting-tree structure allows a single feasible option for the analysis of the neighboring regions: to analyze the data distribution in one predecessor of cell a_h and in the face neighbors of that predecessor. Experiments (see the upcoming Sect. 4.7) show that top clustering accuracy is obtained by choosing the first (direct) predecessor of a_h, cell a_{h-1}, which defines *at most six* regions to analyze per axis. Thus, $Halite_0$ uses this option. Other options are: to analyze 12 regions per axis, by choosing predecessor a_{h-2}; to analyze 24 regions per axis, by choosing predecessor a_{h-3}; and so on. But, these options would force $Halite_0$ to require more than two levels in the tree, the minimum requirement when choosing a_{h-1}, which would make the method slower.

For axis e_j, the neighbor cells to be analyzed are the predecessor a_{h-1} of a_h, its internal neighbor $NI(a_{h-1}, e_j)$ and its external neighbor $NE(a_{h-1}, e_j)$. Together, they have $nP_j = a_{h-1}.n + NI(a_{h-1}, e_j).n + NE(a_{h-1}, e_j).n$ points. The half-space counts in these three cells show how the points are distributed in *at most six* consecutive, equal-sized regions in axis e_j, whose densities $Halite_0$ shall analyze. The point count in the center region, the one containing a_h, is given by: $cP_j = a_{h-1}.P[j]$, if the j-bit in $a_h.loc$ is 0, or by $cP_j = a_{h-1}.n - a_{h-1}.P[j]$ otherwise. For example, regarding cell A_3 in Fig. 4.3b and axis e_1, the six analyzed regions are presented in distinct texture, $cP_1 = 1$ and $nP_1 = 6$.

If at least one axis e_j of cell a_h has cP_j significantly greater than the expected average number of points $\frac{nP_j}{6}$, $Halite_0$ assumes that a new β-cluster was found. Thus, for each axis e_j, the null hypothesis test is applied to compute the probability that the central region contains cP_j points if nP_j points are uniformly distributed in the six analyzed regions. The critical value for the test is a threshold to which cP_j must be compared to determine whether or not it is statistically significant to reject the null hypothesis. The statistic significance is a user-defined probability α of wrongly rejecting the null hypothesis. For a one-sided test, the critical value θ_j^α is computed as $\alpha = Probability(cP_j \geq \theta_j^\alpha)$. The probability is computed assuming the Binomial distribution with the parameters nP_j and $\frac{1}{6}$, since $cP_j \sim Binomial(nP_j, \frac{1}{6})$, under the null hypothesis and $\frac{1}{6}$ is the probability that one point falls into the central region, when it is randomly assigned to one of the six analyzed regions. If $cP_j > \theta_j^\alpha$ for at least one axis e_j, $Halite_0$ assumes a_h to be the center cell of a new β-cluster and increments $_\beta k$. Otherwise, the next tree level is processed.

4.3.2.3 Describing the β-cluster

Once a new β-cluster was found, $Halite_0$ generates the array of relevances $r = [r_1, r_2, \ldots r_d]$, where $r[j]$ is a real value in (0, 100] representing the relevance of axis e_j regarding the β-cluster centered in a_h. The relevance $r[j]$ is given by $(100 * cP_j)/nP_j$. Then, $Halite_0$ automatically tunes a threshold to mark each axis as relevant or irrelevant to the β-cluster. The relevances in $r[\ ]$ are sorted in ascending order into array $o = [o_1, o_2, \ldots o_d]$, which is analyzed to find the best cut position p, $1 \leq p \leq d$ that maximizes the homogeneity of values in the partitions of $o[\ ]$, $[o_1, \ldots o_{p-1}]$ and $[o_p, \ldots o_d]$. The value $cThreshold = o[p]$ defines axis e_j as relevant or irrelevant, by setting $V[\beta k][j] = true$ if $r[j] \geq cThreshold$, and false otherwise.

In order to identify the value of p, based on the MDL principle, $Halite_0$ analyzes the homogeneity of values in partitions of $o[\ ]$, $[o_1, \ldots o_{p'-1}]$ and $[o_{p'}, \ldots o_d]$, for all possible cut positions p', integer values between 1 and d. The idea is to compress each possible partition, representing it by its mean value and the differences of each of its elements to the mean. A partition with high homogeneity tends to allow good compression, since its variance is small and small numbers need less bits to be represented than large ones do. Thus, the best cut position p is the one that creates the partitions that compress best.

For a given p', the number of bits required to represent the respective compressed partitions of $o[\ ]$ is computed using the following equation.

$$size\left(p'\right) \;=\; b\left(\mu_L\right) \;+\; \sum_{1 \leq j < p'} b\left(o[j] - \mu_L\right) + \; b\left(\mu_R\right) \;+\; \sum_{p' \leq j \leq d} b\left(o[j] - \mu_R\right)$$

$$(4.1)$$

In Equation 4.1, $b()$ is a function that returns the number of bits required to represent the value received as input, μ_L is the mean of $[o_1, \ldots o_{p'-1}]$ and μ_R is the mean of $[o_{p'}, \ldots o_d]$. It is assumed $b\left(\mu_L\right) = 0$ for $p' = 1$. Minimizing $size()$, through all possible cut positions p', integer values between 1 and d, leads $Halite_0$ to find the best cut position p.

The last step required to identify the new β-cluster is to find its lower and upper bounds in each axis. These bounds are respectively set to $L[\beta k][j] = 0$ and $U[\beta k][j] = 1$ for every axis e_j, having $V[\beta k][j] = false$. For the other axes, the relevant ones, these bounds are first set equal to the lower and upper bounds of a_h in these axes. Then, they are refined analyzing the neighbors of a_h. Considering a relevant axis e_j, if there exists a non-empty face neighbor of a_h whose lower bound is smaller than the lower bound of a_h, then $L[\beta k][j]$ is decreased by $1/2^h$. In the same way, if there exists a non-empty face neighbor of a_h whose upper bound is bigger than the upper bound of a_h, then $U[\beta k][j]$ is increased by $1/2^h$. In this way, $Halite_0$ completes the description of the β-cluster and restarts applying the mask from level two of the tree to find another β-cluster. The process stops when the mask is applied to every tree level and no new β-cluster is found.

4.3.2.4 Time and Space Complexity

Algorithm 2 identifies $_\beta k$ β-clusters. When looking for each β-cluster, at most $H-2$ tree levels are analyzed, which have at most η cells each. For each tree level, the cells that do not belong to a previously found β-cluster are the convolution pivots to apply the mask. Finally, the neighborhood of the cell with the largest convoluted value is analyzed to find if it is the center of a new β-cluster. Thus, the time complexity of this part of Algorithm 2 (lines 3–18) is $O(_\beta k^2 H^2 \eta d)$. After finding each new β-cluster, the relevance level array with d real values in $(0, 100]$ is built in $O(d)$ time and sorted in $O(d \log d)$ time, the principle MDL is used in $O(d)$ time and the new β-cluster is described in $O(d H)$ time. Thus, the time complexity of this part of Algorithm 2 (lines 19–28) is $O(d_{\beta}k(\log d + H))$. However, each iteration step of the first part of Algorithm 2 consumes a time t_1 that is much larger than the time t_2 consumed by each iteration step of the second part. Thus, the total time of Algorithm 2 is $O(_\beta k^2 H^2 \eta d) t_1 + O(d_{\beta}k(\log d + H)) t_2$. Given that t_1 and t_2 are constant values and $t_1 \gg t_2$, one can say that the method $Halite_0$ is quasi-linear in d, and experiments corroborate this claim. See the upcoming Sect. 4.7 for details. The space complexity of Algorithm 2 is $O(d_{\beta}k + d + H)$, as it builds the matrices L, U and V and it uses arrays with either d or H positions each.

4.3.3 Building the Correlation Clusters

The final phase of $Halite_0$ builds $_\gamma k$ correlation clusters based on the β-clusters found before. According to Definition 4.3, $Halite_0$ analyzes pairs of β-clusters and those that overlap are merged into a single cluster, including merged clusters in further mergers. Algorithm 3 details this phase.

4.3.3.1 Time and Space Complexity

Algorithm 3 analyzes, at cost $O(d_{\beta}k^2)$, all the $_\beta k$ β-clusters found before to identify and to merge, among all possible pairs, the ones that overlap in the d-dimensional space. Then, it defines, at cost $O(d_{\gamma}k_{\beta}k)$, the axes relevant to each of the $_\gamma k$ merged clusters, based on the relevant axes of the $_\beta k$ β-clusters. Thus, the time complexity of Algorithm 3 is $O(d(_\beta k^2 + _\gamma k_{\beta}k))$. During the process, an array with $_\beta k$ positions links β-clusters to correlation clusters, while a matrix with $_\gamma k$ lines and d columns indicates the relevant axes to the clusters. Thus, the space complexity of Algorithm 3 is $O(_\beta k + _\gamma k d)$.

4.4 Presented Method: The Algorithm *Halite*

In this section we describe the *Halite* **method for correlation clustering**. It improves the basic *Halite_0* by providing an optimized implementation strategy for the Counting-tree, even for the case when it does not fit in main memory. The *Halite_0*

Algorithm 3 : Building the correlation clusters.

Input: matrices of β-clusters L, U and V, number of β-clusters $_\beta k$
Output: set of correlation clusters C, number of correlation clusters $_\gamma k$
1: identify all pairs of β-clusters that overlap;
2: merge each pair of overlapping β-clusters into a single cluster, including merged clusters in further mergers;
3: define the points that belong to each merged cluster, as the ones that belong to at least one of its β-clusters;
4: define the dimensions relevant to each merged cluster, as those relevant to at least one of its β-clusters;
5: $C =$ the merged clusters;
6: $_\gamma k =$ the number of merged clusters;

---

algorithm has linear space complexity with regard to the number of points, axes and clusters. However, using the recommended configuration, the amount of memory required by it in experiments (see the upcoming Sect. 4.7 for details) varied between 25 and 50 % of the data size, depending on the points distribution. Thus, for large datasets, the use of Operational System's disk cache may become a considerable bottleneck. In order to overcome this problem, *Halite* has a table-based implementation that never uses disk cache, regardless of the input dataset. Therefore, it allows one to efficiently analyze large amounts of data.

The idea is to represent the Counting-tree by tables stored in main memory and/or in disk. Each table represents one tree level, by storing in key/value entries the data related to all non-empty cells of that level. Remember that $Halite_0$ uses cells with the structure $< loc, n, P[d], usedCell, ptr >$, where loc is the cell spatial position inside its parent cell, n is the number of points in the cell, $P[\,]$ is an array of half-space counts, $usedCell$ is a boolean flag and ptr is a pointer to the next tree level. For *Halite*, this cell structure was slightly modified. Here, the pointer ptr does not exist and loc has the **absolute** position for the cell. In each key/value pair, loc is the key, and the other attributes form the value.

Figure 4.5 exemplifies the data storage for *Halite*. The tables shown consist in a different way of storing the Counting-tree of Fig. 4.3c. Both approaches represent the same data, the 2-dimensional dataset from Fig. 4.3b, the one used in the examples of Sect. 4.3. To reduce cluttering in the figure, the $usedCell$ flags are not shown. Notice, for example, that the cell A_3 from Fig. 4.3b has a single point. This information is stored in both versions of the tree as $A_3.n = 1$. The space position of A_3 is given by $A_3.loc = [11 \downarrow 00 \downarrow 01]$ in Fig. 4.5. This information is found in Fig. 4.3c as $[A_1.loc \downarrow A_2.loc \downarrow A_3.loc] = [11 \downarrow 00 \downarrow 01]$. Finally, the half-space counts are represented by $A_3.P[1] = 1$ and $A_3.P[2] = 0$ in both data structures.

Provided the algorithm $Halite_0$ and the fact that the tree can be stored in tables with key/value entries, the implementation of *Halite* is done as follows: use a traditional approach to store tables with key/value entries in main memory and/or in disk for the tree storage, and apply to *Halite* the same strategies used by $Halite_0$, described in Algorithms 1, 2 and 3. Notice that, between the used data structures, the Counting-tree is the only one that may have large changes in size with regard to the input

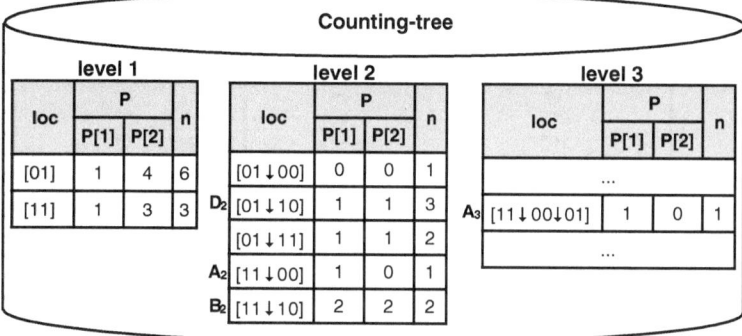

Fig. 4.5 The Counting-tree by tables of key/value pairs in main memory and/or in disk. It allows *Halite* to efficiently cluster large datasets, even when the tree does not fit in main memory. Notice that, both this tree and the one in Fig. 4.3c represent the example dataset from Fig. 4.3b

dataset. Thus, considering this table-based implementation, it is possible to affirm that *Halite* never uses disk cache, regardless of the input dataset.

The current implementation for *Halite* stores the Counting-tree by using the Oracle Berkeley DB 11g[2] configured for simple data storage to avoid data locks. The cache size is set according to the available main memory. It currently has hash tables storing the key/value pairs, but other structures could also be used, such as B-trees.

4.5 Presented Method: Soft Clustering

The algorithm *Halite* is a hard clustering method, i.e., it defines a dataset partition by ensuring that each point belongs to at most one cluster. Hard clustering methods lead to high quality results for most datasets. Also, several applications require the definition of a dataset partition. However, hard clustering is not the best solution for some specific cases, in which the clusters have high probability to overlap. Consider the 2-dimensional dataset in Fig. 4.6a. The data contain a pair of clusters that overlap, making *any* dataset partition not a good choice, since the points in light-gray should belong to both clusters. In cases like that, the so-called soft clustering methods are more appropriate, as they allow points in the overlapping spaces to belong to more than one cluster. For that reason, this section describes the *Halite* **method for soft correlation clustering**, a soft clustering approach for *Halite*.

As a real example, let us consider the clustering analysis of satellite images. In this scenario, a topographer wants to analyze terrains in a set of images, usually assuming that each image is split into tiles (say, 32×32 pixels), from which features are extracted. The topographer expects to find clusters of 'water' tiles, 'concrete'

---

[2] www.oracle.com/technology/products/berkeley-db/

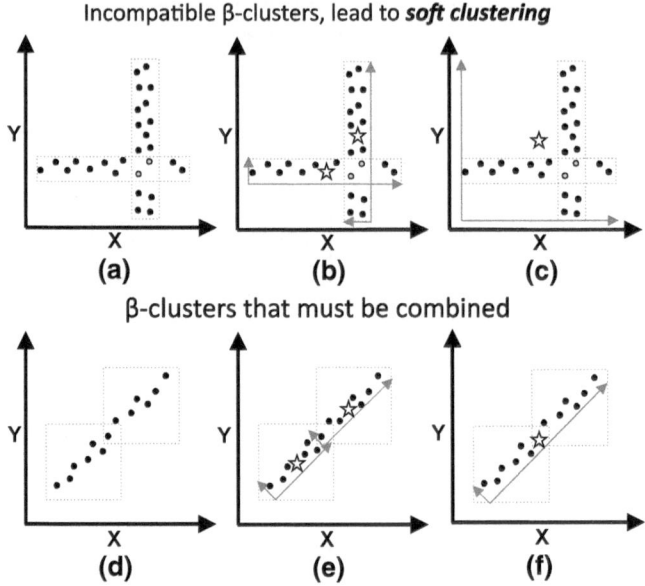

Fig. 4.6 Illustration of the soft clustering method *Halite$_s$*: β-clusters (*dotted rectangles*) may stay apart (*top*) if they are incompatible, resulting in soft clustering (*light-gray circles* in (**a**)); or merged together (*bottom*). The compression-based formulas of *Halite$_s$* automatically make the right choice. In all cases, a '*star*' indicates the center of the respective clusters

tiles, 'forest' tiles, etc., but the used procedure tends to create many hybrid tiles, as a bridge (both 'water' and 'concrete') or a park ('water', 'forest' and 'concrete'), which should belong to more than one cluster. In other words, there is a high probability that the clusters overlap in the data space. Therefore, a hard clustering method is semantically inappropriate to the case.

Halite$_s$ is a fast and scalable algorithm carefully designed to spot points that should belong to two or more clusters, being recommended to scenarios where the probability of cluster overlap is high, as in our previous example with satellite imagery. The algorithm has three phases. The first two are the same ones used for hard clustering: Algorithms 1 and 2, including the improvements presented in Sect. 4.4. The third phase is new. It takes β-clusters as input and uses a compression-based analysis to combine some of the ones that overlap into a soft clustering result. Figure 4.6 illustrates the problem. Distinct 2-dimensional datasets are shown in Fig. 4.6a and d. Each dataset contains a pair of overlapping β-clusters, described by dotted rectangles. The β-clusters in Fig. 4.6a clearly form two clusters and, thus they should remain apart. However, the ones in Fig. 4.6d should be combined into a single cluster. *Halite$_s$* automatically makes the right choice in both cases.

The full pseudo-code for the new phase three is shown in Algorithm 4. The idea is to use the Minimum Description Length (MDL) principle and to analyze the compression ratio of pairs of overlapping β-clusters, where the clusters in each pair can be compressed apart or combined. $Halite_s$ picks the option that compresses best. If the combined cluster allows better compression than the β-clusters do in separate, the β-clusters are merged; otherwise, the β-clusters remain apart, allowing points in the overlapping spaces to belong to both clusters.

Algorithm 4 : Building soft clusters.

Input: matrices of β-clusters L, U and V, number of β-clusters $_\beta k$
Output: set of correlation clusters C, number of correlation clusters $_\gamma k$
1: **for** $k' = 1, 2, ..., _\beta k$ **do**
2:    **for** $k'' = k' + 1, k' + 2, ..., _\beta k$ **do**
3:      **if** β-clusters $_\beta^{\delta'} C_{k'}$ and $_\beta^{\delta''} C_{k''}$ overlap **then**
4:        $size_{k'} = compress(_\beta^{\delta'} S_{k'})$, **if** not done yet;
5:        $size_{k''} = compress(_\beta^{\delta''} S_{k''})$, **if** not done yet;
6:        $size_{k' \cup k''} = compress(_\beta^{\delta'} S_{k'} \cup _\beta^{\delta''} S_{k''})$;
7:        **if** $(size_{k'} + size_{k''}) > size_{k' \cup k''}$ **then**
8:          // the combined cluster compresses best
9:          merge β-clusters $_\beta^{\delta'} C_{k'}$ and $_\beta^{\delta''} C_{k''}$ into a single cluster, including merged clusters in further mergers, as it is done in Algorithm 3;
10:       **end if**
11:     **end if**
12:   **end for**
13: **end for**

The strategy used for compression is described in Algorithm 5, which refers to the function *compress*( ) used in Algorithm 4. The function receives a set of points $_\gamma^\delta S_k$, related to a possible cluster $_\gamma^\delta C_k$ and returns the size of the input data in bits, after compression. Notice: in order to avoid additional disk accesses, $_\gamma^\delta S_k$ is approximated by the tree level where the respective β-clusters were found, assuming points in the center of the cluster's cells. The compression is performed as follows. The principal components of $_\gamma^\delta S_k$ are computed, and the points of $_\gamma^\delta S_k$ are projected into **all** d principal components. Let the projected points be in a set $_\gamma^\delta P_k$ and the d principal components computed be in a set $_\gamma A_k$. Notice that this step states that $_\gamma^\delta P_k$ and $_\gamma A_k$ define the possible cluster $_\gamma^\delta C_k$ as $_\gamma^\delta C_k = \left\langle _\gamma A_k, _\gamma^\delta P_k \right\rangle$ after its projection into the principal components of $_\gamma^\delta S_k$. Then, the maximum max_j and minimum min_j values of $_\gamma^\delta P_k$ in each principal component $a_j \in _\gamma A_k$ are found. After that, one can represent the input data $_\gamma^\delta S_k$ in a compressed way by storing: the descriptions of each principal component a_j, besides their related values for max_j and min_j, and; the differences to the center, $p_{ij} - (((max_j - min_j)/2) + min_j)$, for each projected point $p_i \in _\gamma^\delta P_k$, with regard to each principal component $a_j \in _\gamma A_k$. The size in bits needed to store these items is the output. Large numbers need more bits to be stored than small numbers do. Thus, a set of points in which the stored differences to the center are small tend to lead to a good compression.

Algorithm 5 : Function compress( ).

Input: set of points $\overset{\delta}{\gamma} S_k$ for a possible cluster $\overset{\delta}{\gamma} C_k$

Output: compressed size for $\overset{\delta}{\gamma} S_k$

1: compute the principal components of $\overset{\delta}{\gamma} S_k$;

2: $\overset{\delta}{\gamma} P_k = \overset{\delta}{\gamma} S_k$, projected into **all** d principal components computed;

3: $_\gamma A_k$ = the set of d principal components computed;

   // Notice that Lines 2 and 3 state that $\overset{\delta}{\gamma} P_k$ and $_\gamma A_k$ define the possible cluster $\overset{\delta}{\gamma} C_k$ as $\overset{\delta}{\gamma} C_k = \left(_\gamma A_k, \overset{\delta}{\gamma} P_k \right)$ after its projection into the principal components of $\overset{\delta}{\gamma} S_k$.

4: max_j, min_j = the maximum and the minimum values of $\overset{\delta}{\gamma} P_k$, in each principal component $a_j \in \, _\gamma A_k$;

5: $size$ = the number of bits needed to store the descriptions of each principal component $a_j \in \, _\gamma A_k$, besides its related values for max_j and min_j;

6: **for** every projected point $p_i \in \overset{\delta}{\gamma} P_k$ **do**

7:    **for** $j = 1, 2, ..., d$ **do**

8:       // difference to the center wrt principal component $a_j \in \, _\gamma A_k$

9:       $size = size +$ number of bits needed to store $(p_{ij} - (((max_j - min_j)/2) + min_j))$;

10:    **end for**

11: **end for**

12: **return** size;

Let us illustrate how the compression idea works, by using the example datasets in Fig. 4.6d, Dotted rectangles, gray arrows and stars refer to β-clusters, principal components and cluster centers respectively. First, consider the data in Fig. 4.6a. For the β-clusters apart, Fig. 4.6b shows that we have 'medium compression', since the differences to the centers are small in one principal component of each cluster and large in the other. However, Fig. 4.6c shows that the compression is worse for the combined cluster, since the differences to the center are large in both principal components. Thus, $Halite_s$ keeps the β-clusters apart. Notice that this step defines that the points in gray will belong to both clusters in the final result. For the data in Fig. 4.6, it is respectively shown in Fig. 4.6e and f that the differences to the centers are small in one principal component and large in the other, both with the β-clusters in separate and combined. However, in the first case, $Halite_s$ stores the descriptions of two sets of principal components and the related values of max_j and min_j for each component. A single set is stored for the combined cluster, which leads to better compression. Therefore, $Halite_s$ decides to combine the β-clusters into a single correlation cluster.

4.6 Implementation Discussion

A possible bottleneck in the algorithm described is related to computing the critical value θ_j^α for the statistical test, line 14 of Algorithm 2. As shown in Sect. 4.3, the new algorithm carefully identifies data space regions that refer to bumps in the point density and verifies if these regions stand out in the data in a statistical sense, thus spotting clusters. The Binomial distribution $Binomial(n, p)$ is the base for the statistical test. But, computing the critical value with the exact Binomial distribution may become a bottleneck for large values of n. Fortunately, an efficient approximation to the $Binomial(n, p)$ is given by the Normal distribution $Normal(n.p,\ n.p.(1-p))$. Also, it is common sense in statistics that the approximation quality is excellent when $n.p > 5 \wedge n.(1-p) > 5$. Thus, both $Halite$ and $Halite_s$ compute θ_j^α using the normal approximation to the Binomial distribution whenever this rule applies. The exact computation is very efficient in all other cases.

4.7 Experimental Results

This section presents the experiments performed to test the algorithms described in the chapter. The experiments intend to answer the following questions:

1. Compared with seven of the recent and related works, how good is the clustering method $Halite$?
2. How do the new techniques scale up?
3. How sensitive to the input parameters are the new techniques?
4. What are the effects of soft clustering in data with high probability of cluster overlap and, compared with a well-known, soft clustering algorithm, how good is $Halite_s$?

All experiments were made in a machine with 8.00 GB of RAM using a 2.33 GHZ core. The new methods were tuned with a fixed configuration in all experiments, i.e., default values for $\alpha = 1.0E - 10$ and $H = 4$. The justification for this choice is in the upcoming Sect. 4.7. Finally, notice that results on memory usage are not reported neither for $Halite$ nor for $Halite_s$, since both allow the efficient use of data partially stored in disk, as described in Sect. 4.4. Note however that the experiments do compare the memory needs of the related works with those of $Halite_0$. Moreover, remember that $Halite_0$, $Halite$ and $Halite_s$ have similar space complexity, the difference being that $Halite$ and $Halite_s$ do manage the memory in disk if the Counting-tree does not fit in main memory.

4.7.1 Comparing Hard Clustering Approaches

This section compares $Halite$ with seven of the top related works over synthetic and real data. The techniques are: ORCLUS [2, 3], COPAC [1], CFPC [19], HARP [18], LAC [6], EPCH [13] and P3C [11, 12]. All methods were tuned to find disjoint clusters. The code of ORCLUS was kindly provided by Kevin Y. Yip and the project Biosphere. The source codes for all other methods were kindly provided by their original authors (i.e., Arthur Zimek and Elke Achtert provided COPAC;

Man Lung Yiu and Nikos Mamoulis provided CFPC; Kevin Y. Yip provided HARP; Carlotta Domeniconi provided LAC; Raymond Chi-Wing Wong provided EPCH, and; Gabriela Moise provided P3C.). Results for $Halite_0$ are also reported, which allows one to evaluate the improvements presented on the basic algorithm.

4.7.1.1 Evaluating the Results

The quality of each result given by each technique is measured based on the well-known precision and recall measurements. The evaluation distinguishes between the clusters known to exist in a dataset $^d S$, which are named as real clusters, and those that a technique finds, which are named as found clusters. A real cluster $^{\delta}_r C_k = \left(_r A_k, {^{\delta}_r} S_k \right)$ is defined as a set $_r A_k$ of δ axes, aligned or not to the original axes, together with a set of points $^{\delta}_r S_k$ densely clustered when projected into the subspace formed by $_r A_k$. Notice that the symbol A is used here in place of the symbol E to represent a set of axes, since the axes in $_r A_k$ can be original axes, but they can also be linear combinations of the original axes, i.e., $_r A_k$ is not necessarily a subset of E. A found cluster $^{\delta}_f C_k = \left(_f A_k, {^{\delta}_f} S_k \right)$ follows the same structure of a real cluster, using the symbol f instead of r. Finally, $_f k$ and $_r k$ respectively refer to the numbers of found and real clusters existing in dataset $^d S$.

For each found cluster $^{\delta}_f C_k$, its most dominant real cluster $^{\delta'}_r C_{k'}$ is identified by the following equation.

$$^{\delta'}_r C_{k'} \mid |^{\delta}_f S_k \cap {^{\delta'}_r} S_{k'}| = max(|^{\delta}_f S_k \cap {^{\delta''}_r} S_{k''}|), \ 1 \le k'' \le {_r}k \qquad (4.2)$$

Similarly, for each real cluster $^{\delta'}_r C_{k'}$, its most dominant found cluster $^{\delta}_f C_k$ is identified by the equation as follows.

$$^{\delta}_f C_k \mid |^{\delta'}_r S_{k'} \cap {^{\delta}_f} S_k| = max(|^{\delta'}_r S_{k'} \cap {^{\delta''}_f} S_{k''}|), \ 1 \le k'' \le {_f}k \qquad (4.3)$$

The precision and the recall between a found cluster $^{\delta}_f C_k$ and a real cluster $^{\delta'}_r C_{k'}$ are computed as follows.

$$precision = \frac{\left| ^{\delta}_f S_k \cap {^{\delta'}_r} S_{k'} \right|}{\left| ^{\delta}_f S_k \right|} \qquad (4.4)$$

$$recall = \frac{\left| ^{\delta}_f S_k \cap {^{\delta'}_r} S_{k'} \right|}{\left| ^{\delta'}_r S_{k'} \right|} \qquad (4.5)$$

To evaluate the quality of a clustering result, one averages the *precision* (Eq. 4.4) for all found clusters and their respective most dominant real clusters. Also, one averages the *recall* (Eq. 4.5) for all real clusters and their respective most dominant found clusters. These two averaged values are closely related to well-known measurements. The first one is directly proportional to the dominant ratio [2, 13], while the second one is directly proportional to the coverage ratio [13]. The harmonic mean of these averaged values is named as *Quality*. The evaluation of a clustering result with regard to the quality of the subspaces uncovered is similar. One also computes the harmonic mean of the averaged *precision* for all found clusters and the averaged *recall* for all real clusters, but exchanging the sets of points (S sets) in the two last equations, Eqs. 4.4 and 4.5, by sets of axes (A sets). This harmonic mean is named as *Subspaces Quality*.

Finally, in the cases where a technique does not find clusters in a dataset, the value zero is assumed for both qualities.

4.7.1.2 System Configuration

Halite uses fixed input parameter values, as defined in the upcoming Sect. 4.7. *Halite$_0$* was tuned in the same way. The other algorithms were tuned as follows. ORCLUS, LAC, EPCH, CFPC and HARP received as input the number of clusters present in each dataset. Also, the known percent of noise for each dataset was informed to HARP. The extra parameters of the previous works were tuned as in their original authors' instructions. LAC was tested with integer values from 1 to 11, for the parameter $1/h$. However, its run time differed considerably with distinct values of $1/h$. Thus, a time out of three hours was specified for LAC executions. All configurations that exceeded this time limit were interrupted. EPCH was tuned with integer values from 1 to 5 for the dimensionalities of its histograms and several real values varying from 0 to 1 were tried for the outliers threshold. For the tests in P3C, the values $1.0E - 1$, $1.0E - 2$, $1.0E - 3$, $1.0E - 4$, $1.0E - 5$, $1.0E - 7$, $1.0E - 10$ and $1.0E - 15$ were tried for the *Poisson threshold*. HARP was tested with the Conga line data structure. CFPC was tuned with values $5, 10, 15, 20, 25, 30$ and 35 for w, values $0.05, 0.10, 0.15, 0.20$ and 0.25 for α, values $0.15, 0.20, 0.25, 0.30$ and 0.35 for β and 50 for *maxout*. ORCLUS was tested with its default values for $\alpha = 0.5$ and $k_0 = 15k$, where k is the known number of clusters present in each dataset. It also received as input the known average cluster dimensionality of each synthetic dataset, and all possible dimensionalities were tested for the real data. COPAC was tuned with its default values for $\alpha = 0.85$ and $k = 3d$ and its default distance function was used. Its parameter ε was defined as suggested in COPAC's original publication and μ received the smallest value between k and the known size of the smallest cluster present in each dataset, since COPAC demands $\mu \leq k$.

Notice two remarks: (a) each non-deterministic previous work was ran 5 times in each possible configuration and the results were averaged. The averaged values were taken as the final result for each configuration; (b) all results reported for the previous methods refer to the configurations that led to the best Quality value, over all possible parameters tuning.

4.7.1.3 Synthetic Data Generation

Synthetic data were created following standard procedures used by most methods described in Chap. 3, including the tested methods. The details are in Algorithm 6. In a nutshell: (1) the used procedure initially created axes-aligned, elliptical clusters of random sizes that follow normal distributions with random means and random variances in at least 50 % of the axes (relevant axes), spreading through at most 15 % of these axes domains. In other axes, the irrelevant ones, all clusters follow the uniform distribution, spreading through the whole axes domains; and (2) an optional data rotation allowed creating clusters not aligned to the original axes. In this step, each dataset was rotated four times in random planes and random degrees.

Algorithm 6 was used to create synthetic datasets organized in several groups. A first group of datasets was created to analyze each tested method with regard to increasing numbers of points, axes and clusters. It contains 7 non-rotated datasets with d, η and $_\gamma k$ growing together from 6 to 18, 12 to $120k$ and 2 to 17 respectively. Noise percentile was fixed at 15 %. For identification purposes, the datasets are named

---

Algorithm 6 : Function generate_one_dataset( ).

Input: dimensionality d, cardinality η, number of clusters $_\gamma k$, percent of noise pN,
   choice between axes-aligned and arbitrarily oriented clusters *rotate*

Output: one synthetic dataset $^d S$

1: // generates the noise points
   $^d S = \eta * (pN/100)$ random points in $[0, 1)^d$;
2: // defines the sizes of the clusters
   $c[ ]$ = array of $_\gamma k$ random integers, such that $\sum_k c[k] = \eta * ((100 - pN)/100)$;
3: **for** $k = 1, 2, ...,_\gamma k$ **do**
4:     // empty matrix (new cluster)
      define $cluster[c[k]][d]$;
5:     // $\delta \geq 50\%$ of d
      δ = random integer in $[\frac{d}{2}, d]$;
6:     // relevant axes, at least 50 %
      randomly pick δ axes;
7:     **for** $j = 1, 2, ..., d$ **do**
8:        **if** axis e_j was picked **then**
9:           // the cluster will spread through at most 15 % of the domain of e_j
            choose random *mean* and *variance*, such that the *Normal(mean, variance)* distribution
            generates values in $[0, 1)$ that differ at most 0.15 from each other;
10:          $cluster[:][j] = c[k]$ real numbers following *Normal(mean, variance)*;
11:       **else** $cluster[:][j] = c[k]$ random values in $[0, 1)$;
12:       **end if**
13:    **end for**
14:    insert each row of $cluster[ ][ ]$ as a point in $^d S$;
15: **end for**
16: **if** *rotate* **then**
17:    // optional step of rotation
      rotate $^d S$ four times in random planes and in random degrees, and then normalize $^d S$ to $[0, 1)^d$;
18: **end if**
19: **return** $^d S$;

6, 8, 10, 12, 14, 16 and 18d according to their dimensionalities. Rotated versions of these datasets were also created to analyze each method over clusters not aligned to the original axes. These datasets are named 6, 8, 10, 12, 14, 16 and 18d_r.

The strategy employed to create the 14d dataset (14 axes, 90k points, 17 clusters, 15 % noise and non-rotated) was the base for the creation of the other groups of datasets. Based on it, 3 groups of datasets were created varying one of these characteristics: numbers of points, axes or clusters. Each group has datasets, created by Algorithm 6, in which a single characteristic changes, while all others remain the same as the ones in the 14d dataset. The number of points grows from 50 to 250k, the dimensionality grows from 5 to 30 and the number of clusters grows from 5 to 25. The names Xk, Xc and Xd_s refer respectively to datasets in the groups changing numbers of points, clusters or axes. For example, dataset 30d_s differs from 14d only because it has 30 axes instead of 14.

4.7.1.4 Results on Synthetic Data

This section compares the methods on synthetic data. Results for clustering quality, memory consumption and run time are shown. For easy reading the graphs of the section, the values obtained for each method were linked and all vertical axes related to run time or to memory consumption were plotted in *log scale*. When a method found no cluster in a dataset, despite it was ran several times for each possible parameter configuration, the respective values measured for run time and also for memory consumption were ignored, as in most cases each run led to different values measured due to the distinct configurations used. When a method used disk cache, its run time was ignored too. In these cases, no lines link the values obtained for the respective methods in the graphs.

Figure 4.7 presents the results for run time and for clustering accuracy. Figure 4.7a shows that *Halite*, *Halite*$_0$, EPCH, HARP and LAC presented similar high Quality values for all the datasets in the first group. CFPC presented a clear decrease in Quality when the dimensionality was higher than 12. P3C, ORCLUS and COPAC had worse Quality results. Regarding run time, Fig. 4.7b shows that *Halite* was in general the fastest algorithm, loosing by little to *Halite*$_0$ for small data. As an example, for the biggest dataset 18d, *Halite* was respectively 6, 17, 54, 145, 515, 1, 440 and 10, 255 times faster than *Halite*$_0$, CFPC, EPCH, LAC, P3C, ORCLUS and COPAC.

The results for data with individual changes in numbers of points, axes and clusters are shown in Fig. 4.7, from 4.7c–h. Notice that, *Halite*, *Halite*$_0$, LAC and EPCH performed well in Quality for all cases, exchanging positions but being in general within 10 % from each other with no one being prevalent. Notice also that *Halite* and *Halite*$_0$ always showed the same Quality. ORCLUS, COPAC, CFPC, HARP and P3C performed worse than that. *Halite* was again the fastest method in almost all cases, only tying with *Halite*$_0$ for low dimensional datasets. As an example, for the dataset with the highest dimensionality, 30d_s, *Halite* ran respectively 9, 25, 36, 419, 1, 542, 3, 742 and 5, 891 times faster than *Halite*$_0$, CFPC, LAC, P3C, ORCLUS, HARP and COPAC.

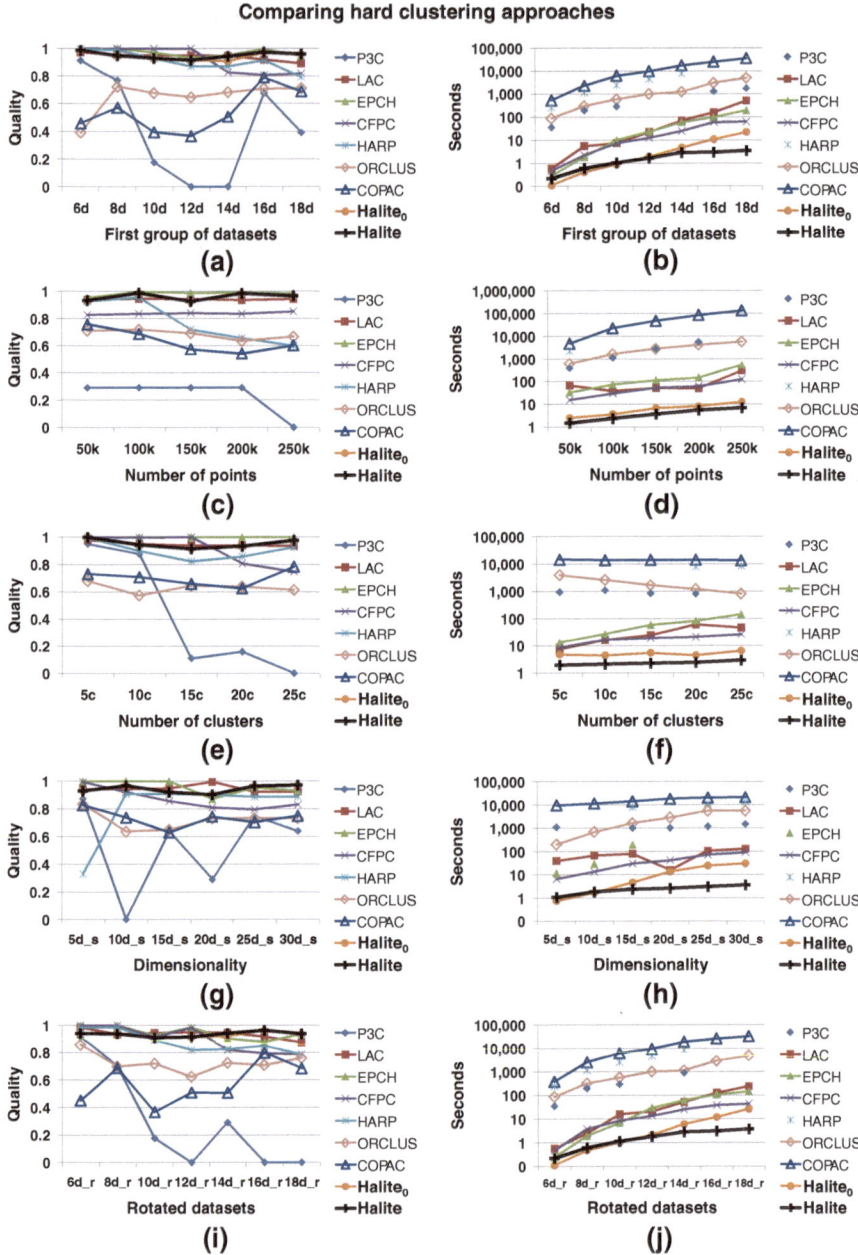

Fig. 4.7 *Halite* is shown in *black vertical crossing lines*. *Left column*: quality; *right column*: wall-clock time in *log scale*. Comparison of approaches for hard clustering - *Halite* was in average at least 12 times faster than seven top related works, always providing high quality clusters

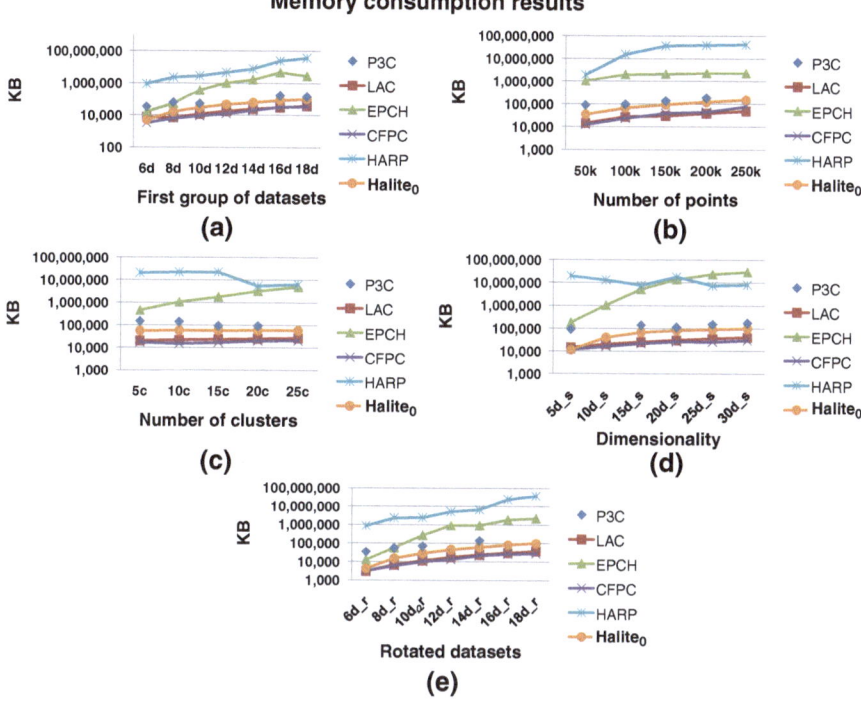

Fig. 4.8 Results on memory consumption for synthetic data

Another experiment refers to the rotated data. It analyzes each method's abilities to find clusters in subspaces formed by linear combinations of original axes. The results are in Fig. 4.7i and j. *Halite*, $Halite_0$ and LAC were only marginally affected by rotation, varying at most 5 % in their respective Quality values, compared to the results of the same data without rotation. All others had considerable decreased or increased Quality values for at least one case. Run time results were similar to those obtained for non-rotated data.

The results on memory usage are presented in Fig. 4.8. Results for P3C, LAC, EPCH, CFPC, HARP, and also for the method $Halite_0$ are reported. Figure 4.8 shows that, in all cases, there was a huge discrepancy between HARP and EPCH face to the others with regard to memory usage. Note the *log scale* in every Y-axis. As an example, for dataset $18d$, the biggest one into the first group of datasets, HARP used approximately 34.4 GB of memory and EPCH used 7.7 % of this amount, while $Halite_0$ used only 0.3 % of the memory required by HARP.

The quality of relevant axes was also evaluated. LAC, COPAC and ORCLUS were not tested here, as they do not return a set of original axes to define the axes relevant to each cluster. The results for the first group of datasets are in Fig. 4.9. The Subspaces Quality values are similar for *Halite*, $Halite_0$ and EPCH. All others had worse results. The same pattern was seen in the other datasets.

Fig. 4.9 Subspace quality for synthetic data

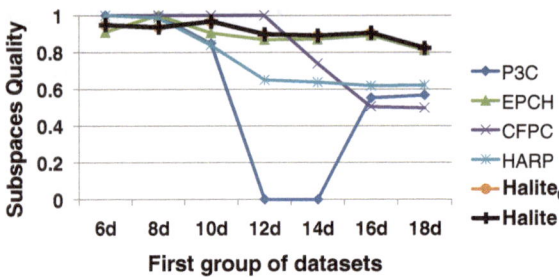

Concluding, P3C had the worst Quality values in most cases. HARP, CFPC, ORCLUS and COPAC provided average Quality values for some datasets, but the results were not good in several cases. *Halite*, *Halite$_0$*, LAC and EPCH had the best results, in general tying with regard to Quality values. However, contrasting to *Halite* and also to *Halite$_0$*, the methods LAC and EPCH demanded guessing the number of clusters and required distinct threshold tuning to each dataset to obtain their best results reported. Regarding memory needs, remember that *Halite$_0$*, *Halite* and *Halite$_s$* have similar space complexity, the difference being that *Halite* and *Halite$_s$* do manage the memory in disk if the Counting-tree does not fit in main memory. With that in mind, the memory needs of the related works were compared with those of *Halite$_0$*. This comparison shows that CFPC in general required the least amount of memory, followed by LAC, *Halite$_0$*, P3C, EPCH and HARP which respectively required 1.2, 2.8, 6.5, 112 and 600 times more memory than CFPC in average. Therefore, the memory consumption of *Halite$_0$* was similar to those of top related works. Regarding run time, *Halite* was the fastest method in almost all cases, only loosing by little to *Halite$_0$* in some small datasets. The improvements on the basic implementation allowed *Halite* to be about one order of magnitude faster than *Halite$_0$* for large data. *Halite* also avoids the use of disk cache. Finally, when comparing to the related works, *Halite* was in average 12, 26, 32, 475, 756, 2, 704 and 7, 218 times faster than CFPC, LAC, EPCH, P3C, ORCLUS, HARP and COPAC respectively. Notice that *Halite* was in average at least 12 times faster than seven recent and related works, always providing high quality clusters.

4.7.1.5 Real Data

The real dataset used to test the methods is the training data provided for the Siemens KDD Cup 2008.[3] It was created for automatic breast cancer diagnosis, consisting of 25 of the most significant features extracted automatically from 102,294 Regions of Interest (ROIs) present in X-ray breast images of 118 malignant cases and 1,594 normal cases. Ground truth is also provided. The data was partitioned into four datasets, each containing features extracted from homogeneous images, i.e., each

---

[3] "http://www.kddcup2008.com"

Fig. 4.10 Quality versus run time in *linear-log scale* over 25-dimensional data for breast cancer diagnosis (KDD Cup 2008). *Halite* was at least 11 times faster than 5 previous works (2 other failed), increasing their accuracy in up to 35 %. Similar behavior occurred in synthetic data

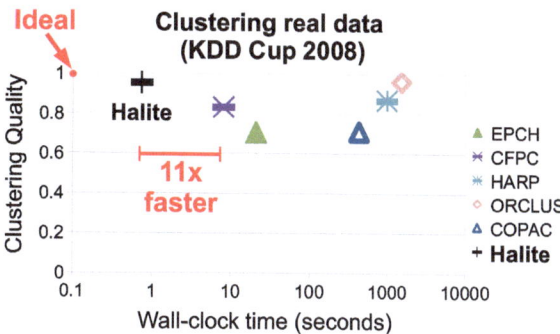

dataset has features extracted from $\sim 25,000$ ROIs related to images taken from one breast, left or right, in one of the possible directions, CC or MLO. The Quality results were computed based on the ground truth class label of each ROI.

4.7.1.6 Results on Real Data

All methods were tested with the real data. However, LAC and P3C failed for all four datasets in all tested configurations. LAC always grouped all points into a single cluster. P3C did not finish within a week for all cases. Thus, they are not reported. The results for left breast images in one MLO view are shown in Fig. 4.10. *Halite* was at least 11 times faster than the previous works, increasing their accuracy in up to 35 %. The other three real datasets led to similar results.

4.7.2 Comparing Soft Clustering Approaches

This section compares *Halite$_s$* with STATPC [10], a well-known, state-of-the-art soft clustering method. The original code for STATPC was used, which was gracefully provided by Gabriela Moise. The input parameter values used for STATPC were the default values suggested by its authors, $\alpha_0 = 1.0E - 10$ and $\alpha_K = \alpha_H = 0.001$. *Halite$_s$* uses fixed input parameter values, as defined in the upcoming Sect. 4.7.4. The methods were compared in a scenario with high probability of cluster overlap, the example scenario with satellite imagery from Sect. 4.5. 14 high quality satellite images from famous cities were analyzed, as the city of Hong Kong in Fig. 4.11a. The images, available at "geoeye.com", have a combined size of 17 MB. Each image was divided into equal-sized rectangular tiles, from which Haar wavelets features were extracted. The process led to a 10-dimensional dataset of 14, 336 points.

Figure 4.11b and c respectively exemplify the results for STATPC and for *Halite$_s$* over this data by coloring each tile from the example image of Hong Kong according to its cluster. As expected, some tiles belong to more than one cluster. These were

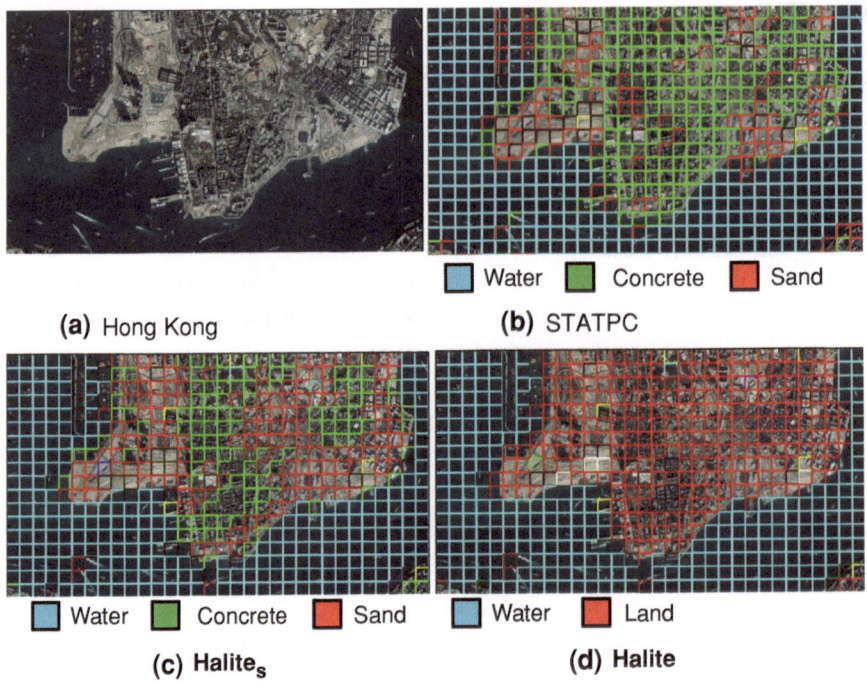

(a) Hong Kong   **(b)** STATPC

█ Water   █ Concrete   █ Sand

█ Water   █ Concrete   █ Sand   █ Water   █ Land

(c) Halite$_s$ **(d) Halite**

Fig. 4.11 Comparing *Halite* and *Halite$_s$* with STATPC on data with cluster overlap. As expected, hard clustering leads to correct, but less detailed results: roughly speaking, *Halite$_s$* and STATPC found clusters of 'water' tiles (*cyan*), 'sand' tiles (*red*) and 'concrete' tiles (*green*); *Halite* merged the last two clusters into a cluster of 'land' tiles (*red*). The results for *Halite$_s$* and STATPC are similar. Notice, however, that *Halite$_s$* found the clusters in only *two seconds*, while STATPC took *two days* to perform the same task. Therefore, the new solution can be used in *real time applications*. IKONOS/GeoEye-1 Satellite image courtesy of GeoEye

colored according to their first clusters assigned. Notice that both results are similar, with clusters that represent the main patterns apparent in the example image. However, STATPC took *two days* to find the clusters, while *Halite$_s$* performed the same task in only *two seconds*. Similar results were obtained for the other 13 images. Notice that these results indicate that the new method allows the development of *real time applications*, like a software to aid on the fly the diagnosis process in a worldwide Healthcare Information System or a system to look for deforestation within the Amazon Rainforest in real time.

Finally, results for *Halite* (Fig. 4.11d) are reported, which, as expected, are still correct, but provide less details compared to the soft clustering ones. Roughly speaking, *Halite$_s$* and STATPC found clusters of 'water' tiles (cyan), 'sand' tiles (red) and 'concrete' tiles (green), whereas *Halite* merged the last two clusters into a single cluster of 'land' tiles (red). These results corroborate the conjecture from Sect. 4.5, that soft clustering is more appropriate to this kind of data.

Fig. 4.12 Scalability of
Halite and *Halite$_s$* on synthetic data of varying sizes
and dimensionality. Plots in
log-log scale. Notice: both
methods scale as expected,
according to the theoretical
complexity analysis from
Sect. 4.3

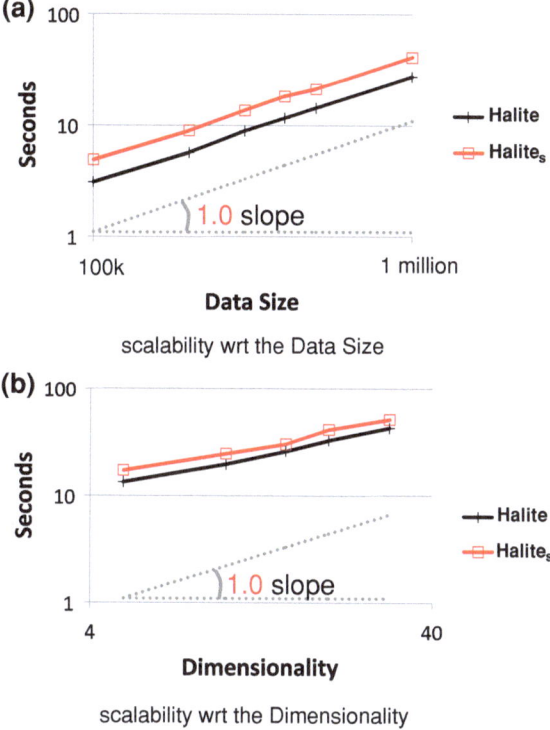

4.7.3 Scalability

This section analyzes the scalability of *Halite* and *Halite$_s$* with regard to increasing data sizes and dimensionality. Synthetic datasets were created by Algorithm 6.
The data sizes and dimensionality vary from $100\,k$ to 1 *million* points and from 5 to
30 axes respectively. The datasets have 10 axes-aligned clusters each and 15 % of
noise. Notice in Fig. 4.12 that the new techniques scale as expected, according to the
theoretical complexity analysis, presented in Sect. 4.3.

4.7.4 Sensitivity Analysis

The behavior of the new techniques varies based on two parameters: α and H. This
section analyzes how they affect these methods. Both parameters were varied for
Halite, *Halite$_0$* and *Halite$_s$* to maximize the Quality values obtained, defining the best
configuration for each dataset and technique. Then, for each dataset and technique,
the best configuration was modified, changing one parameter at a time, and the
technique's behavior was analyzed. For example, when varying H for a dataset and

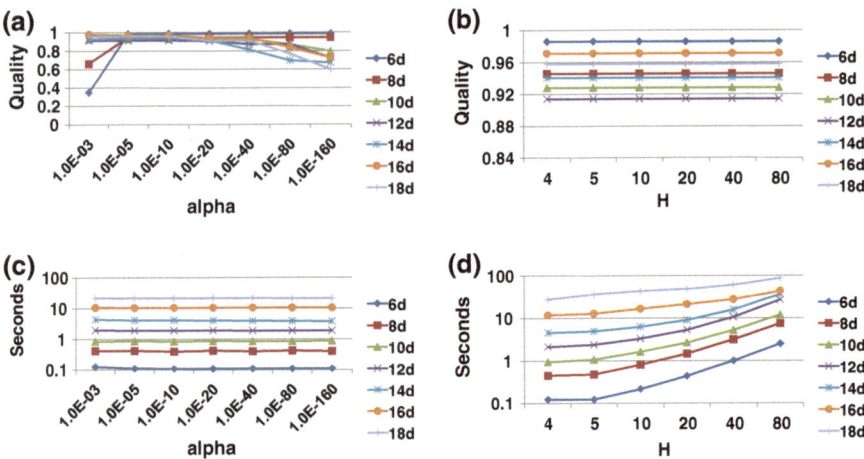

Fig. 4.13 Sensitivity analysis that defined the default configuration, $\alpha = 1E - 10$ and $H = 4$

technique, the value of α was fixed at the value in the respective best configuration. The tested values of α and H vary from $1.0E - 3$ to $1.0E - 160$ and from 4 to 80 respectively. Figure 4.13 reports the results of $Halite_0$. Figure 4.13a and c present the results regarding α. The values of α that led to the best Quality vary from $1.0E - 5$ to $1.0E - 20$ and the run time was barely affected by changes in α. Concerning H, Fig. 4.13b and d show that the Quality does not increase significantly for H higher than 4. However, the run time increased as expected regarding H. Thus, small values for H, such as 4, are enough for most datasets. Similar results were obtained by $Halite_0$, $Halite$ and $Halite_s$ for all synthetic and real datasets. Therefore, the values $\alpha = 1.0E - 10$ and $H = 4$ are considered the default configuration for the new techniques. These fixed values were used in all experiments.

4.8 Discussion

This section provides a discussion on some specific characteristics of the new clustering methods described. As a first topic to discuss, notice that the new method looks for dense space regions, and thus it works for any data distribution. This fact was illustrated on rotated Gaussians, as well as on real data of unknown distribution.

The quadratic behavior on the number of β-clusters found is not a crucial concern for the new techniques. Experimental evaluation showed that this number closely follows the ground truth number of clusters existing in the tested datasets. In the experiments, the biggest number of β-clusters found over all synthetic and real datasets was 33. Notice that, the biggest number of clusters present in these data is 25. The results for the first group of synthetic datasets are in Fig. 4.14, which shows a plot with the Y-axis referring to each dataset tested, and horizontal bars representing the

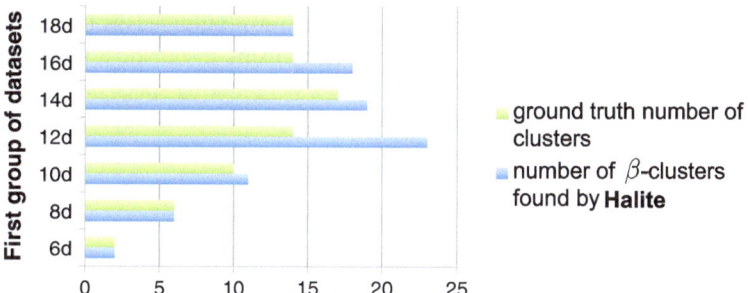

Fig. 4.14 Ground truth number of clusters versus the number of β-clusters found by *Halite* over synthetic data

respective number of β-clusters found by *Halite*, besides the ground-truth number of clusters. The same pattern was seen in all other datasets analyzed. Furthermore, the analysis of data with many clusters is usually meaningless, as it is hard to the user to obtain semantic interpretations from a large number of clusters.

Additionally, the quadratic behavior on H is not a relevant concern for *Halite*, since very small values for H are commonly sufficient to obtain accurate clustering results. Remember that a Counting-tree describes a d-dimensional dataset in H distinct resolutions. Each cell in the first resolution level $h = 0$ is divided in 2^d cells in level $h + 1$, which are divided again in 2^d cells in level $h + 2$ each, and so on. The process stops when $h = H - 1$. Thus, for moderate-to-high dimensionality data, the maximum count of points in a cell of tree level h converges exponentially to 1, as the value of h increases to reach $H - 1$. After the point of convergence, even for a skewed dataset, more levels tend to be useless, since they would not help to better describe the data. The sensitivity analysis in Sect. 4.7 corroborate this claim.

Also, notice that, *Halite* is limited to the size of the clusters that it finds. The method analyses the points distribution in specific regions of the data space with all dimensions using an statistical hypothesis test to identify β-clusters, which lead to the clusters. But, these regions must have a minimum amount of points to reject the null hypothesis. In this way, *Halite* may miss clusters with small amount of points present in low-dimensional subspaces, as they tend to be extremely sparse in spaces with several dimensions. On the other hand, the clustering results tend to be better as the number of points in the clusters increase. Thus, *Halite* is suitable for large, multi-dimensional datasets and as it scales linearly on the data size, it becomes better as the dataset increases. In this way, *Halite* tends to be able to spot accurate clusters even from datasets with more than 30 axes, when they are large enough.

Notice one last remark: the *traditional* clustering method STING [17] is a basis to the work described in this chapter. Similarly to *Halite*, STING also does multi-resolution space division in a statistical approach for clustering. However, STING is a *traditional* clustering method that proposes to only analyze two or very low dimensional data. It is *not* suitable for moderate-to-high dimensionality data clustering, since STING does *not* spot clusters that only exist in subspaces of the original data space. Also, STING uses a fixed density threshold to find clusters, whereas *Halite* applies a novel spike detection strategy based on convolution masks to find possible clusters, and then *Halite* confirms the clusters by using a statistical test to identify the spikes that are significantly denser than their neighboring space regions. Finally, as opposed to Halite, STING does not include a soft clustering approach.

4.9 Conclusions

This chapter described the new **method** *Halite* **for correlation clustering**. Previous methods are typically super-linear in space or execution time. The main strengths of *Halite* are that it is fast and scalable, while still giving highly accurate results. In details, the main contributions of *Halite* are:

1. **Scalability**: it is linear in time and space with regard to the data size and to the dimensionality of the clusters. *Halite* is also linear in memory usage and quasi-linear in running time regarding the space dimensionality;
2. **Usability**: it is deterministic, robust to noise, does not have the number of clusters as a parameter and finds clusters in subspaces formed by original axes or by their linear combinations, including space rotation;
3. **Effectiveness**: it is accurate, providing results with equal or better quality compared to top related works;
4. **Generality**: it includes a soft clustering approach, which allows points to be part of two or more clusters that overlap in the data space. Specifically, the new algorithm encompasses: *Halite$_0$*, a basic implementation for the clustering method; *Halite*, the optimized, and finally recommended method for hard clustering, and; *Halite$_s$*, the recommended implementation for soft clustering.

A theoretical study on the time and space complexity of *Halite*, shown in Sect. 4.3, as well as an extensive experimental evaluation performed over synthetic and real data spanning up to 1 *million* elements corroborate these properties. Specifically, the experiments compared *Halite* with seven representative works. On synthetic data, *Halite* was consistently the fastest method, always presenting highly accurate results. Regarding real data, *Halite* analyzed 25-dimensional data for breast cancer diagnosis (KDD Cup 2008) at least 11 times faster than five previous works, increasing their accuracy in up to 35 %, while the last two related works failed.

Halite is the first algorithm described in this book that focuses on data mining in large sets of complex data. The next chapters describe two other algorithms aimed at tackling this difficult problem.

References

1. Achtert, E., Böhm, C., Kriegel, H.P., Kröger, P., Zimek, A.: Robust, complete, and efficient correlation clustering.In: SDM, USA (2007).

2. Aggarwal, C., Yu, P.: Redefining clustering for high-dimensional applications.IEEE TKDE 14(2), 210–225 (2002). doi:10.1109/69.991713
3. Aggarwal, C.C., Yu, P.S.: Finding generalized projected clusters in high dimensional spaces. SIGMOD Rec. 29(2), 70–81 (2000). doi:10.1145/335191.335383
4. Cordeiro, R.L.F., Traina, A.J.M., Faloutsos, C., Traina Jr., C.: Finding clusters in subspaces of very large, multi-dimensional datasets. In: F. Li, M.M. Moro, S. Ghandeharizadeh, J.R. Haritsa, G. Weikum, M.J. Carey, F. Casati, E.Y. Chang, I. Manolescu, S. Mehrotra, U. Dayal, V.J. Tsotras (eds.) ICDE, pp. 625–636. IEEE (2010).
5. Cordeiro, R.L.F., Traina, A.J.M., Faloutsos, C., Traina Jr., C.: Halite: Fast and scalable multi-resolution local-correlation clustering. IEEE Transactions on Knowledge and Data Engineering 99(PrePrints) (2011). doi:10.1109/TKDE.2011.176.16 pages
6. Domeniconi, C., Gunopulos, D., Ma, S., Yan, B., Al-Razgan, M., Papadopoulos, D.: Locally adaptive metrics for clustering high dimensional data. Data Min. Knowl. Discov. 14(1), 63–97 (2007). doi:10.1007/s10618-006-0060-8
7. Gonzalez, R.C., Woods, R.E.: Digital Image Processing, 3rd edn. Prentice-Hall, Inc., Upper Saddle River, NJ, USA (2006)
8. Grunwald, P.D., Myung, I.J., Pitt, M.A.: Advances in Minimum Description Length: Theory and Applications (Neural Information Processing). The MIT Press (2005).
9. Jolliffe, I.T.: Principal Component Analysis, 2nd edn. Springer-Verlag, New York, USA (2002)
10. Moise, G., Sander, J.: Finding non-redundant, statistically significant regions in high dimensional data: a novel approach to projected and subspace clustering. In: KDD, pp. 533–541 (2008).
11. Moise, G., Sander, J., Ester, M.: P3C: A robust projected clustering algorithm. In: ICDM, pp. 414–425. IEEE Computer Society (2006).
12. Moise, G., Sander, J., Ester, M.: Robust projected clustering. Knowl. Inf. Syst. 14(3), 273–298 (2008). doi:10.1007/s10115-007-0090-6
13. Ng, E.K.K., chee Fu, A.W., Wong, R.C.W.: Projective clustering by histograms. TKDE 17(3), 369–383 (2005). doi:10.1109/TKDE.2005.47
14. Rissanen, J.: Stochastic Complexity in Statistical Inquiry Theory. World Scientific Publishing Co., Inc., River Edge, NJ, USA (1989)
15. Traina Jr, C., Traina, A.J.M., Faloutsos, C., Seeger, B.: Fast indexing and visualization of metric data sets using slim-trees. IEEE Trans. Knowl. Data Eng. **14**(2), 244–260 (2002)
16. Traina Jr., C., Traina, A.J.M., Wu, L., Faloutsos, C.: Fast feature selection using fractal dimension. In: SBBD, pp. 158–171 (2000).
17. Wang, W., Yang, J., Muntz, R.: Sting: A statistical information grid approach to spatial data mining.In: VLDB, pp. 186–195 (1997).
18. Yip, K., Cheung, D., Ng, M.: Harp: a practical projected clustering algorithm. TKDE **16**(11), 1387–1397 (2004). doi:10.1109/TKDE.2004.74
19. Yiu, M.L., Mamoulis, N.: Iterative projected clustering by subspace mining. TKDE 17(2), 176–189 (2005). doi:10.1109/TKDE.2005.29.

Chapter 5
BoW

Abstract The large amounts of data collected by enterprises are accumulating data, and today it is already feasible to have Terabyte- or even Petabyte-scale datasets that must be submitted for data mining processes. However, given a *Terabyte-scale* dataset of moderate-to-high dimensionality, how could one cluster its points? Numerous successful, serial clustering algorithms for data in five or more dimensions exist in literature, including the algorithm *Halite* that we described in the previous chapter. However, the existing algorithms are impractical for datasets spanning Terabytes and Petabytes, and examples of applications with such huge amounts of data in five or more dimensions abound (e.g., Twitter crawl: >12 TB, Yahoo! operational data: 5 *Petabytes* [6]). This limitation was previously summarized in Table 3.1. For datasets that do not even fit on a single disk, parallelism is a first class option, and thus we must re-think, re-design and re-implement existing serial algorithms in order to allow for parallel processing. With that in mind, this chapter presents one work that explores parallelism using MapReduce for clustering huge datasets. Specifically, we describe in detail one second algorithm, named *BoW* [5], that focuses on data mining in large sets of complex data.

Keywords Correlation clustering · Terabyte-scale data mining · MapReduce · Hadoop · Big data · Complex data · Large graphs · Social networks

5.1 Introduction

Given a *Terabyte-scale* dataset of moderate-to-high dimensional elements, how could one cluster them? Numerous successful, serial clustering algorithms for data in five or more dimensions exist in literature, including the algorithm *Halite* that we described in the previous chapter. However, the existing algorithms are impractical for data spanning Terabytes and Petabytes (e.g., Twitter crawl: >12 TB, Yahoo! operational data: 5 *Petabytes* [6]). In such cases, the data are *already* stored on multiple disks, as the largest modern disks are 1–2 TB. Just to read a single Terabyte of data (at 5 GB/min on a single modern eSATA disk) one takes more than 3 hours. Thus, parallelism is not another option—it is by far the best choice. Nevertheless, good, serial clustering algorithms and strategies are still extremely valuable, because we can (and should) use them as 'plug-ins' for parallel clustering. Naturally, the best algorithm is the one that combines (a) a fast, scalable serial algorithm and (b) makes it run efficiently in parallel. This is exactly what the method described here does.

Examples of applications with Terabytes of data in five or more dimensions abound: weather monitoring systems and climate change models, where we want to record wind speed, temperature, rain, humidity, pollutants, etc; social networks like Facebook TM, with millions of nodes, and several attributes per node (gender, age, number of friends, etc); astrophysics data, such as the Sloan Digital Sky Survey, with billions of galaxies and attributes like red-shift, diameter, spectrum, etc.

This chapter focuses on the problem of finding clusters in subspaces of *Terabyte-scale, moderate-to-high dimensionality* datasets. The method described here uses `MapReduce`, and can treat as a plug-in almost any of the serial clustering methods, including the algorithm *Halite* described in the previous chapter. The major research challenges addressed are (a) how to minimize the I/O cost, taking into account the *already existing* data partition (e.g., on disks), and (b) how to minimize the network cost among processing nodes. Either of them may be the bottleneck. Thus, we present the ***Best of both Worlds***—*BoW* method [5] that automatically spots the bottleneck and chooses a good strategy. The main contributions of *BoW* are as follows:

1. *Algorithm design and analysis:* the method *BoW* is based on a novel, adaptive algorithm that automatically picks the best of two strategies and good parameters for it, hence its name ***Best of both Worlds***. One of the strategies uses a novel *sampling-and-ignore* idea that reduces the network traffic;
2. *Effectiveness, scalability and generality: BoW* can use most serial clustering methods as plug-ins, including the method *Halite* described before. *BoW* requires no user defined parameters (due to its defaults) and it maintains the serial clustering quality, with near-linear scale-up;
3. *Experiments:* experiments on real and synthetic data with *billions* of elements were performed, using hundreds of machines running in parallel.

Experiments were performed on a combination of large real and synthetic datasets, including the *Yahoo! web* one.[1] To the best of our knowledge, the Yahoo! web is the largest real dataset for which results have ever been reported in the database clustering literature for data in five or more axes. Although spanning 0.2 TB of multi-dimensional data, *BoW* took only 8 min to cluster it, using 128 cores. The experiments also used up to 1,024 cores, again the highest such number reported in the clustering literature for moderate-to-high dimensional data.

Notice one *important* remark: *BoW* is *tailored to spot clusters in subspaces* of moderate-to-high dimensionality data and it can handle most serial algorithms as plug-ins, since the only required API is that the serial algorithm should return clusters

---

[1] Provided by Yahoo! Research (www.yahoo.com).

of points in hyper-rectangles, which we shall refer to as β-clusters in this book, whose definition follows the same one previously employed for the *Halite* algorithm, but which may also be provided by many other existing algorithms. Overlapping β-clusters are then merged to form clusters. Indeed, the intuition is to generalize the structure of isometric crystal systems to the d-dimensional case in order to describe clusters of any shape and size, existing in subspaces only, as it was extensively discussed in the previous Chap. 4. Remember that the clustering methods well-suited to analyze moderate-to-high dimensionality data spot clusters that exist only in subspaces of the original, d-dimensional space (i.e., spaces formed by subsets of the original axes or linear combinations of them). Thus, the *natural* shape of the clusters in the original space facilitates their representation with hyper-rectangles, as the points of each cluster spread linearly through several axes (original axes or linear combinations of them) in the original space. For that reason, many of the existing serial, clustering methods (e.g., CLIQUE [1, 2], FPC/CFPC [10], P3C [8, 9], STATPC [7], and *Halite* [3, 4]) return clusters in hyper-rectangles, and adapting others to work with *BoW* tends to be facilitated by the natural shape of the clusters. Nevertheless, besides focusing on spotting clusters in subspaces of moderate-to-high dimensionality data, *BoW* also works with traditional clustering methods and low dimensional data, if the plug-in returns clusters in hyper-rectangles.

5.2 Main Ideas of BoW: Reducing Bottlenecks

The major research problems for clustering *very large* datasets with `MapReduce` are (a) how to minimize the I/O cost, and (b) how to minimize the network cost among processing nodes. *Should we split the data points at random, across machines? What should each node do, and how should we combine the results? Do we lose accuracy (if any), compared to a serial algorithm on a huge-memory machine?*

The method described here answers all of those questions, by careful design and by adaptively trading-off disk delay and network delay. Specifically, we describe a novel, adaptive algorithm named *BoW* that is a hybrid between two strategies presented in this section: (1) the *ParC* method that does data partitioning and merges the results; and (2) the *SnI* method that does some sampling first, to reduce communication cost at the expense of higher I/O cost. There is no universal winner between *ParC* and *SnI*, since it depends on the environment used and also on the dataset characteristics (see the upcoming Sect. 5.5 for details). *BoW* automatically picks the best option, and good parameters for it. The reason for the success of *BoW* is its upcoming cost-estimation formulas (Eqs. 5.4 and 5.5), which help *BoW* to pick the best alternative and to set proper parameters for the chosen environment, while requiring nimble computational effort. Next, we describe the methods *ParC* and *SnI* in detail.

5.2.1 Parallel Clustering: ParC

The *ParC* algorithm has three steps: (1) appropriately partition the input data and assign each data partition to one machine, (2) each machine finds clusters in its assigned partition, named as β-clusters, and, (3) merge the β-clusters found to get the final clusters. There are subtle issues on how to merge the results once clustering is done on each machine, which are detailed in the upcoming Sect. 5.4.

We give the details in the upcoming Sect. 5.4, but in a nutshell, three options for data partitioning are considered: (a) *random data partitioning*: elements are assigned to machines at random, striving for load balance; (b) *address-space data partitioning*: eventually, nearby elements in the data space often end up in the same machine, trading-off load balance to achieve better merging of the β-clusters; and (c) *arrival order or 'file-based' data partitioning*: the first several elements in the collection go to one machine, the next batch goes to the second, and so on, achieving perfect load balance. The rationale is that it may *also* facilitate the merging of the β-clusters, because data elements that are stored consecutively on the disk, may be nearby in address space too, due to locality: For example, galaxy records from the Sloan Digital Sky Survey (SDSS) are scanned every night with smooth moves of the telescope, and thus galaxies close in (2-dimensional) address space, often result in records that are stored in nearby locations on the disk.

As described in Sect. 2.5, a `MapReduce`-based application has at least two modules: the map and the reduce. The *ParC* method partitions the data through `MapReduce` mappers and does the clustering in `MapReduce` reducers. The final merging is performed serially, since it only processes the clusters descriptions, which consist of a tiny amount of data and processing. Figure 5.1a (Fig. 5.1b will be explained latter) illustrates the process. It starts in phase **P1** with m mappers reading the data in parallel from the `MapReduce` distributed file system. In this phase, each mapper receives a data element at a time, computes its key, according to the data partition strategy used, and outputs a pair $\langle key, point \rangle$. All elements with the same key are forwarded in phase **P2** to be processed together, by the same reducer, and the elements with distinct keys are processed apart, by distinct reducers.

In phase **P3**, each reducer receives its assigned set of elements and normalizes them to a unitary hyper-cube. Each reducer then applies the plugged-in serial clustering algorithm over the normalized elements, aiming to spot β-clusters. For each β-cluster found, the reducer outputs, in phase **P4**, a pair $\langle key, cluster\_description \rangle$. The key concatenates the reducer identification and a cluster identification. The reducer identification is the input key. The cluster identification is a sequential number according to the order in which the β-cluster was found in the corresponding reducer. A β-cluster description refers to the unnormalized minimum/maximum limits of the cluster in each axis, defining a hyper-rectangle in the data space. Notice that this is a tiny amount of data, amounting to two float values per axis, per β-cluster.

The last step is phase **P5**, that involves merging and/or stitching the β-clusters provided by all the reducers to calculate the final answer. This step is performed serially, as it processes only the tiny amount of data (the bounds of each β-cluster found)

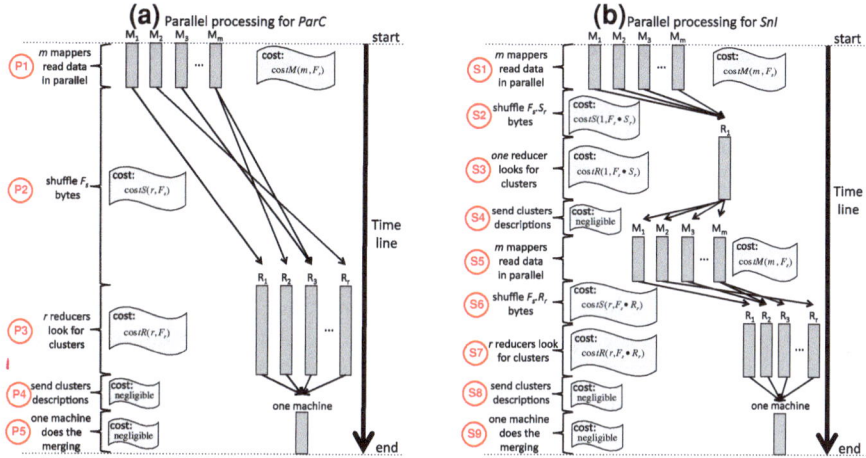

Fig. 5.1 *Which one is best?* Parallel run overview for *ParC* (*left*) and *SnI* (*right*—with sampling). *ParC* executes the map (*P1*), shuffle (*P2*) and reduce (*P3*) phases once, on the full dataset. *SnI* uses sampling (phases *S1–S4*) to get rough cluster estimates and then uses phases *S5–S9* to cluster the remaining points (see Sect. 5.2.2 for details). Their clustering accuracies are similar (see the upcoming Sect. 5.5). The winning approach depends on the environment; *BoW* uses cost-based optimization to automatically pick the best.

received from phase **P4**, and *not* the data elements themselves. The best strategy to follow in this step is highly dependent on the criteria used by the mapper to partition the data. Thus, *ParC* uses distinct procedures for distinct data partitioning criteria. The procedures used for each of the partitioning strategies studied are detailed in the upcoming Sect. 5.4.

5.2.2 Sample and Ignore: SnI

The initial implementation for parallel clustering, the *ParC* algorithm, reads the dataset once aimed at minimizing disk accesses, which is the most common strategy used by serial algorithms to shrink computational costs. However, this strategy does not address the issue of minimizing the network traffic: in the shuffle phase of the *ParC* algorithm (phase **P2** of Fig. 5.1a) almost all of the records have to be shipped over the network to the appropriate reducer. It may become a considerable bottleneck. *How can this network traffic be reduced?*

The main idea in this section is to minimize the network traffic for parallel clustering by exploiting the skewed distribution of cluster sizes that typically appears in real data. Most of the elements usually belong to a few large clusters, and these are exactly the elements that we try to *avoid* processing. Thus, we describe *SnI*, a parallel clustering algorithm that consists of: (a) a novel *sample-and-ignore* preprocessing step; and (b) the *ParC* algorithm from Sect. 5.2.1. The *sample-and-ignore* step works on a small dataset sample, spots the major clusters and ignores these clusters' members in the follow-up steps. It significantly reduces the amount of data moved in the shuffling phases of *SnI*, with consequent savings for the network traffic, as well as the I/O cost for the intermediate results and processing cost for the receiving reduce tasks. Notice that this *sample-and-ignore* idea is an alternative, general strategy that can improve many clustering methods, and not only *ParC*.

The *SnI* method is defined in Algorithm 7 and the process is illustrated in Fig. 5.1b. At a high-level, in Phase I (steps **S1–S4** in the figure, and lines **1–3** in the algorithm) the method *samples* the input data and builds an initial set of clusters. In the second phase (steps **S5–S9** in the figure, and lines **4–8** in the algorithm), the input data is filtered, so that only *unclassified* elements are included, that is, those that do not belong to any of the clusters found in Phase I. These unclassified elements are then clustered using *ParC*.

Algorithm 7 : Multi-phase Sample-and-Ignore (SnI) Method.

Input: dataset $^d S$, sampling ratio S_r
Output: *clusters*
1: // Phase 1 – Sample
2: m mappers read data and send elements to one reducer with probability S_r;
3: one reducer uses plug-in to find clusters in $\sim \eta.S_r$ received elements, and passes clusters descriptions to m mappers;
4: // Phase 2 – Ignore
5: m mappers read data, ignore elements from clusters found in sample and send the rest to r reducers, according to the Data Partition Approach;
6: r reducers use plug-in to find clusters in the received elements, and send clusters descriptions to one machine;
7: one machine merges clusters received and the ones from sample, let the merged result be *clusters*;
8: **return** *clusters*

Figure 5.2 illustrates the *SnI* approach over a toy dataset, assuming that we have $r = 2$ reducers available for parallel processing. The top part of the figure shows Phase-I. First, in Phase-I (a) the input dataset is read in parallel by m map tasks, each mapper passes the input elements to the same reducer with some probability, for example, 0.5 for the case shown in the figure. A single reducer builds clusters using the sample elements in Phase-I (b). In the example scenario two clusters were found and they are denoted by the gray boxes around the elements. The summary descriptors of the clusters found in Phase-I, i.e., the minimum and maximum limits of the clusters with regard to each dimension, are passed to Phase-II.

In Phase-II (a), m mappers perform a second pass over the data, this time filtering out points that fall in the clusters found in Phase-I, which are denoted by the black boxes. The elements that do not fall into clusters are passed to the two reducers available, as shown in Phase-II (b) and (c), in which we assume that the used partitioning strategy divided the elements into 'black points' and 'white points'. Each reducer

Phase I – sampling

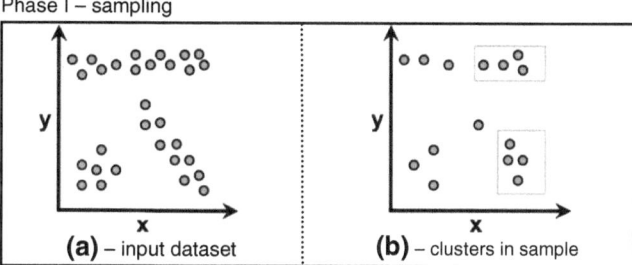

Phase II – look for the clusters not found in the sample

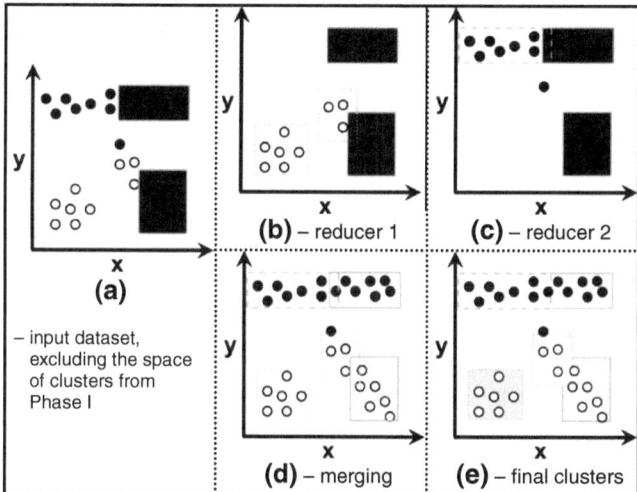

Fig. 5.2 Overview of the multi-phase sample-and-ignore (SnI) method. *Phase-I* finds clusters on a sample of the input data. *Phase-II* ignores elements that fall within a previously found cluster and finds clusters using the remaining elements only.

finds new clusters, denoted by the points surrounded by dotted boxes. In Phase-II (d), the clusters found by the reducers are merged with the clusters from the sampling phase using the upcoming merging/stitching strategies described in Sect. 5.4. The global set of clusters, containing three clusters represented in Phase-II (e) by distinct gray levels, is the final output.

The main benefit of the *SnI* approach is realized in the shuffle/reduce stages. In Phases **S2** and **S3** of Fig. 5.1b, only a small sample is shuffled and processed by a receiving reducer. In Phases **S6** and **S7** of Fig. 5.1b, only the *non-ignored* elements may need to be shuffled through the network to other machines and processed. This means that most elements belonging to the major clusters spotted in the sample are ignored, *never* being shuffled through the network *nor* processed by a reducer. Compared to the *ParC* algorithm, *SnI* significantly minimizes the network cost and the reducers processing, at the cost of reading the whole dataset twice. In other words,

ParC does a single pass over the data, but almost all of the records have to be shipped over the network (in phase **P2** of Fig. 5.1a) to be processed by the appropriate reducer. On the other hand, *SnI* minimizes the shuffle/reduce cost, at the expense of reading the dataset one extra time. *What approach is the best?* The answer is given in Sect. 5.3.

5.3 Cost-Based Optimization of BoW

This section presents an adaptive, hybrid method named *BoW* (*B*est of both *W*orlds) that exploits the advantages of the previously described approaches, *ParC* and *SnI*, taking the best of them. There is no universal winner, since it depends on the environment and on the dataset characteristics. See the upcoming Sect. 5.5 for a complete explanation. Therefore, the main open question is: *When should the novel sampling-and-ignore idea be used and when should it be avoided?* *ParC* executes the map, shuffle and reduce phases only once on the whole dataset. *SnI* reduces the amount of data to be shipped to and processed by the reducers, at the expense of a second pass on the input data (in the map phase). This section presents a cost-based optimization that uses simple analytics models to estimate the running time of each clustering strategy. *BoW* picks the strategy with the lowest estimated cost.

The environmental parameters required by *BoW* are presented in Table 5.1. They describe the hardware characteristics (i.e., the specs of the available `MapReduce` cluster), the total amount of data to be processed, and the cost estimate for the plugged-in serial clustering method. Setting the value for F_s is straightforward. D_s, N_s and $start\_up\_cost(t)$ are inferred by analyzing the cloud of computers' logs, while $plug\_in\_cost(s)$ is defined based on the plugged-in method's original time complexity analysis and/or experiments, or measured by the user in a simple experiment. Notice: each machine in the cloud may run many `MapReduce` tasks (mappers and/or reducers) in parallel, *sharing* the machine's disks and network connection. Therefore, N_s and D_s are expected to be *smaller* than the effective network bandwidth and disk transfer rate respectively.

Two other parameters are used, shown in Table 5.2, and reasonable default values are provided for them based on empirical evaluation. Notice one *important* remark: As is the common knowledge in database query optimization, at the cross-over point of two strategies, the wall-clock-time performances usually create *flat plateaus*, being not much sensitive to parameter variations. This occurs in *BoW*'s setting, and the results in the upcoming Figs. 5.9a, 5.10a and d exemplify it (notice the log–log scale). Thus, tuning exact values to these parameters barely affects *BoW*'s results and the suggested values are expected to work well in most cases.

The following lemmas and proofs define the equations of the cost-based optimization. First, we describe the expected costs for complete map, shuffle and reduce phases relative to the number of mappers and/or reducers available and to the amount of data involved. Then, we show *BoW*'s infered costs for: *ParC*, which minimizes disk accesses, and; *SnI*, which aims at shrinking the network cost. For clarity, consider

Table 5.1 Environmental parameters

Parameter	Meaning	Explanation
F_s	data file size (bytes)	Size of the dataset to be clustered
D_s	disk speed (bytes/sec)	Average number of bytes per second that a MapReduce task (mapper or reducer) is able to read from local disks, i.e. the average disk transfer rate per MapReduce task
N_s	network speed (bytes/sec)	Average bytes/sec. that a MapReduce task (mapper or reducer) is able to read from other computers in the cloud, i.e. the average network transfer rate per MapReduce task
$start\_up\_cost(t)$	start-up cost (seconds)	Average time to start-up t MapReduce tasks (mappers or reducers)
$plug\_in\_cost(s)$	plug-in cost (seconds)	Average time to run the plugged-in serial method over s data bytes on a standard computer in the cloud

again Fig. 5.1 that provides a graphical overview of the parallel execution of both methods, including their expected cost equations.

Lemma 5.1 *Map Cost - the expected cost for the map phase of the parallel clustering approaches is a function of the number of mappers m used and the involved data size s, given by:*

$$cost M(m, s) = start\_up\_cost(m) + \frac{s}{m} \cdot \frac{1}{D_s} \qquad (5.1)$$

Proof In the map phase, m mappers are started-up at the cost of $start\_up\_cost(m)$. Additionally, the majority of the time spent is related to reading the input dataset

Table 5.2 Other parameters

Parameter	Meaning	Explanation	Default values
D_r	dispersion ratio	Ratio of data transferred in the shuffle phase through the network (distinct machines) relative to the total amount of data processed	0.5
R_r	reduction ratio	Ratio of data that do not belong to the major clusters found in the sampling phase of *SnI* relative to the full data size F_s	0.1

from disk. s bytes of data will be read in parallel by m mappers, which are able to read D_s bytes per second each. Thus, the total reading time is given by: $\frac{s}{m} \cdot \frac{1}{D_s}$. $\square$

Lemma 5.2 *Shuffle Cost—the expected shuffle cost of the parallel clustering approach is a function of the number of reducers r to receive the data and the amount of data to be shuffled s, which is given by:*

$$cost\,S(r, s) = \frac{s \cdot D_r}{r} \cdot \frac{1}{N_s} \qquad (5.2)$$

Proof The majority of the shuffling cost is related to shipping data between distinct machines through the network. Whenever possible, MapReduce minimizes this cost by assigning reduce tasks to the machines that already have the required data in local disks. D_r is the ratio of data actually shipped between distinct machines relative to the total amount of data processed. Thus, the total amount of data to be shipped is $s \cdot D_r$ bytes. The data will be received in parallel by r reducers, each one receiving in average N_s bytes per second. Thus, the total cost is given by: $\frac{s \cdot D_r}{r} \cdot \frac{1}{N_s}$. $\square$

Lemma 5.3 *Reduce Cost—the expected cost for the reduce phase is a function of the number of reducers r used for parallel processing and the size s of the data involved, which is:*

$$cost\,R(r, s) = start\_up\_cost(r) + \frac{s}{r} \cdot \frac{1}{D_s} + plug\_in\_cost(\frac{s}{r}) \qquad (5.3)$$

Proof In the reduce phase, r reducers are started-up at cost $start\_up\_cost(r)$. Then, the reducers read from disk s bytes in parallel at the individual cost of D_s bytes per second. Thus, the total reading time is $\frac{s}{r} \cdot \frac{1}{D_s}$. Finally, the plugged-in serial clustering method is executed in parallel over partitions of the data, whose average sizes are $\frac{s}{r}$. Therefore, the approximate clustering cost is $plug\_in\_cost(\frac{s}{r})$. $\square$

Lemma 5.4 *ParC Cost—the expected cost of the ParC algorithm is given by:*

$$cost\,C = cost\,M(m, F_s) + cost\,S(r, F_s) + cost\,R(r, F_s) \qquad (5.4)$$

Proof The parallel processing for *ParC* is as follows: (1) F_s bytes of data are processed in the map phase, by m mappers; (2) F_s bytes of data are shuffled to r reducers in the shuffling phase; (3) F_s bytes of data are processed in the reduce phase by r reducers, and; (4) a single machine merges all the β-clusters found. The last step has a negligible cost, since it performs simple computations over data amounting to two float values per β-cluster, per dimension. Thus, summing the costs of the three initial phases leads to the expected cost for *ParC*. $\square$

Lemma 5.5 *SnI Cost—the expected cost for the SnI algorithm is given by:*

$$costCs = 2 \cdot costM(m, F_s) + costS(1, F_s \cdot S_r) + costR(1, F_s \cdot S_r) +$$
$$costS(r, F_s \cdot R_r) + costR(r, F_s \cdot R_r) \tag{5.5}$$

Proof *SnI* runs two complete map, shuffle and reduce phases. In both map phases, the full dataset is processed by m mappers, at combined cost: $2 \cdot costM(m, F_s)$. In the first shuffle phase, a data sample of size $F_s \cdot S_r$ bytes is shuffled to a single reducer, at cost $costS(1, F_s \cdot S_r)$. The reduce cost to process this sample is: $costR(1, F_s \cdot S_r)$. R_r is the ratio of data that does not belong to the major clusters, the ones found in the sampling phase, relative to F_s. That is, $F_s \cdot (1 - R_r)$ bytes are *ignored* in the Second Phase of *SnI*, while $F_s \cdot R_r$ bytes of data are *not ignored*, being processed after clustering the sample. Both second shuffle and reduce phases involve r reducers. Thus, their combined costs are: $costS(r, F_s \cdot R_r) + costR(r, F_s \cdot R_r)$. The costs for shipping and processing β-clusters descriptions is negligible, since the involved amount of data and processing is tiny.   □

Remark: when the parallel clustering algorithms are executed, the number of distinct key values to be sorted by the MapReduce framework is extremely small; it is *always* the number r of reducers used only. Each reducer handles a single key, so it does not need to do sorting. Thus, the sorting cost is negligible for these approaches. The I/O and network costs are the real bottlenecks. The wall-clock time results measured in all experiments performed (see the upcoming Sect. 5.5) confirm this assertion.

Algorithm 8 describes the main steps of *BoW*. In summary, *ParC* executes the map, shuffle and reduce phases once, involving the full dataset. *SnI* runs these phases twice, but involving less data. *What is the fastest approach?* It depends on the environment used. *BoW* takes the environment description as input and applies cost-based optimization to automatically choose the fastest, prior to the real execution. Provided that the clustering accuracies are similar for both approaches (see the upcoming Sect. 5.5 for a complete explanation), *BoW* actually picks the *'Best of both Worlds'*.

Algorithm 8 : The *Best of both Worlds – BoW* Method.

Input: dataset $^d S$, environmental parameters (Table 5.1), other parameters (Table 5.2),
       number of reducers r, number of mappers m, sampling ratio S_r
Output: *clusters*
1: compute $costC$ from Equation 5.4;
2: compute $costCs$ from Equation 5.5;
3: **if** $costC > costCs$ **then**
4:    // use the *sampling-and-ignore* idea
     clusters = result of *SnI* over $^d S$;
5: **else**
6:    // no sampling
     clusters = result of *ParC* over $^d S$;
7: **end if**
8: **return** *clusters*

5.4 Finishing Touches: Data Partitioning and Cluster Stitching

This section describes three reasonable approaches studied for data partitioning and consequent merging and/or stitching of the clusters found in each partition. Notice that *BoW* works with any of the three partitioning approaches described and, *potentially*, works with any user-defined partitioning strategy.

5.4.1 Random-Based Data Partition

The first alternative is the **Random-Based Data Partition.** Mappers randomly assign data elements to reducers, striving for load balance. Each reducer receives a random sample of the dataset, looks for β-clusters on it, and reports the β-clusters it finds, in terms of their MBRs (Minimum Bounding Rectangles).

The final step merges every pair of β-clusters that overlap in the data space. Notice that, to spot an overlap, *only* the descriptions of the β-clusters (MBRs) are needed, and *not* the elements themselves. Two clusters overlap if they overlap in every axis j. Let u_{ij} and l_{ij} represent respectively the upper and lower bounds of cluster i at axis j. Similarly, let $u_{i'j}$ and $l_{i'j}$ represent the bounds of cluster i' at axis j. Two β-clusters i and i' overlap if $u_{ij} \geq l_{i'j} \ \wedge \ l_{ij} \leq u_{i'j}$ holds for every axis j.

Figure 5.3I illustrates a simulation of this process assuming that we have $r = 2$ reducers. The first reducer gets the points indicated as 'white-circles' and the second one gets the 'black-circles'; both reducers run a typical clustering algorithm, returning the MBRs (Minimum Bounding Rectangles) of the β-clusters they discover (Fig. 5.3I, b and c). Then, *BoW* merges the overlapping β-clusters (Fig. 5.3I, d), and returns the results, indicated as the shaded areas of Fig. 5.3I, e. Notice that some data points may be left as outliers, which is a possibility for all the parallel methods that we describe, as well as for most serial clustering algorithms.

5.4.2 Location-Based Data Partition

The second alternative is the **Location-Based Data Partition.** The idea here is to divide the *address space*, trading-off load balance to achieve better merging of the β-clusters. Specifically, *BoW* partitions the address space into r disjoint regions (say, hyper-rectangles, by bi-secting some coordinate axes), where r is the number of reducers. The mappers are given the boundaries of every region, and direct each element accordingly. In the current implementation, *BoW* has r to be a power of two, since the partitions are created by dividing each dimension in half as needed.

Figure 5.3II illustrates a simulation of the process, using the same toy dataset of Figure 5.3I that we used to illustrate the previous approach. The data elements are assigned to reducers according to their location (vertical dashed line). Again, each of the two reducers generates MBRs of the β-clusters it finds. Then *BoW* (a) merges

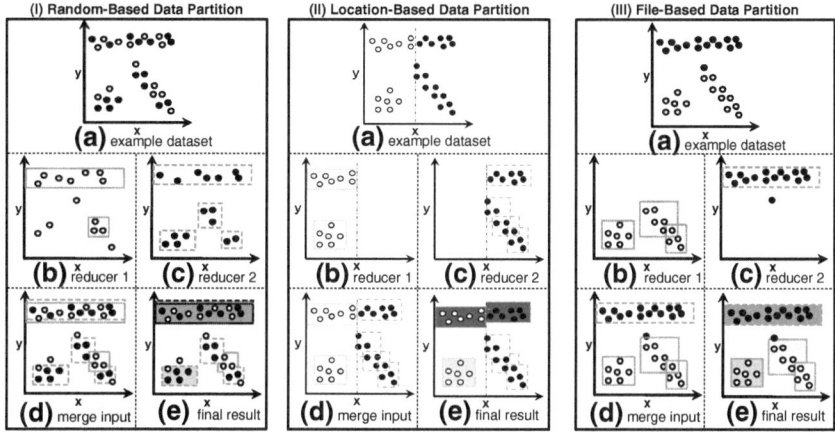

Fig. 5.3 'file-based' wins. Clustering examples for the three data partitioning approaches. We assume exactly the same 2-dimensional input dataset, with $r = 2$ reducers. *I–left* assigns elements at random to reducers, and merges the resulting β-clusters that overlap. *II–middle* divides the address space in disjoint regions, assigns each region to a reducer, and then either merges, or *stitches* the appropriate β-clusters (see Sect. 5.4.2 for details). *III–right*: assigns elements to reducers according to their position in the data file, and hopes that, due to locality, the resulting β-clusters will have little overlap. As shown in the upcoming Sect. 5.5, the 'file-based' strategy outperforms the first two alternatives.

those that overlap and (b) stitches those that touch, like the two β-clusters on the top of Fig. 5.3II, d.

The stitching step requires a careful design. It intends to stitch together the clusters that touch in partitioned positions with respect to one or more axes, and have "enough touching area" with regard to all other axes. In our running example, Fig. 5.4 shows the input for this step. The β-clusters i and i' touch in a partitioned position of axis x. *BoW* proposes to stitch two β-clusters if the area that they jointly touch is larger than the disjoint areas. In more detail, for the example of Fig. 5.4, the "touching area" of the β-clusters i and i' regarding axis y is $h_{i \cap i'}$. As in the illustration, let h_i and $h_{i'}$ be the individual 1-dimensional heights with regard to axis y of the cluster i and i', respectively. *BoW* considers this "touching area" as being "large enough" for stitching if the common part is larger than the union of the non-common parts, for each axis that do not touch in a partitioned position. It is defined by the following equation.

$$h_{i \cap i'} > (h_i - h_{i \cap i'}) + (h_{i'} - h_{i \cap i'}) \tag{5.6}$$

Notice that the "large enough" criterion is *parameter-free*. Algorithm 9 gives the full pseudo-code. In our running example, Fig. 5.3II, e shows the final output for the merging / stitching procedure, assuming that the upper two β-clusters were stitched. The intermediate set of six β-clusters is summarized into three clusters, represented in three distinct gray levels in the illustration.

Algorithm 9 : Stitching β-clusters i and i'.

Input: u_{ij} and l_{ij}, upper and lower bounds of β-cluster i in each axis j
       $u_{i'j}$ and $l_{i'j}$, upper and lower bounds of β-cluster i' in each axis j
Output: merge
1: merge = true;
2: **for** each axis j **do**
3:    **if** (**not** ($u_{ij} \geq l_{i'j} \wedge l_{ij} \leq u_{i'j}$)) $\wedge$ (**not** (axis j was partitioned $\wedge$
       ($u_{ij} = l_{i'j} = partitioned\_position \vee l_{ij} = u_{i'j} = partitioned\_position$))) **then**
4:      // do not overlap neither touch in a partitioned position in j
5:      merge = false;
6:    **end if**
7: **end for**
8: **if** merge **then**
9:    **for** each axis j **do**
10:     **if** ($u_{ij} \geq l_{i'j} \wedge l_{ij} \leq u_{i'j}$) **then**
11:      compute h_i, $h_{i'}$ and $h_{i \cap i'}$ wrt j;
12:      **if** $h_{i \cap i'} \leq (h_i - h_{i \cap i'}) + (h_{i'} - h_{i \cap i'})$ **then**
13:        merge = false; // not "enough touching area" in j
14:      **end if**
15:     **end if**
16:    **end for**
17: **end if**
18: **return** merge

Fig. 5.4 Merging and Stitching for the Location-based approach. *Merging*: the three lower-right β-clusters are merged, since they overlap. *Stitching*: the β-clusters i and i' are stitched to form a bigger cluster, since the height of the "touching area is large enough" compared to the heights of the β-clusters.

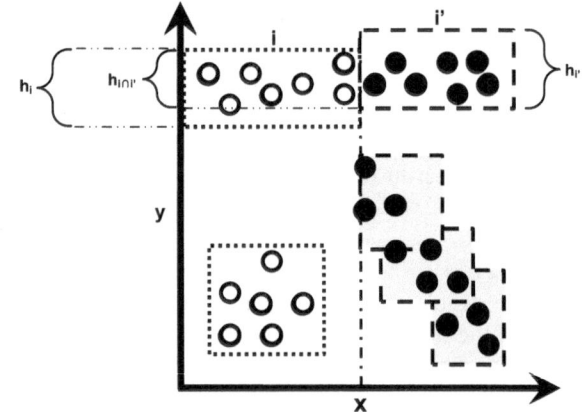

5.4.3 File-Based Data Partition

The third approach is the **File-Based Data Partition.** This approach has perfect load balance, assigning the first $1/r$ portion of the records to the first reducer, the second $1/r$ portion to the second one, and so one. The rationale is that it may *also* facilitate the merging of the β-clusters, because data elements that are stored consecutively on the disk, may be nearby in address space too, due to locality.

The specific steps are as follows: *BoW* intends to divide the input file into r pieces of nearly equal sizes, whose elements are sequentially stored in the file. The `MapReduce` mappers receive the total number of elements η and the total number of reducers r available for parallel processing as input parameters. When an element is received, a mapper takes into account the physical order o of the element in the input file to define its appropriate key. The key is computed by the following equation: $floor(o/ceil((\eta + 1)/r))$, assuring an even amount of elements to each partition. Thus, each reducer receives a set of elements sequentially stored in the input file, and then looks for β-clusters on it. The final step of the computation is identical to the random-based data partitioning approach: *BoW* merges every pair of β-clusters that overlap in the address space.

Figure 5.3III illustrates a simulation of the process assuming that we have $r = 2$ reducers. It follows the same process as in the random-based approach, except for the first step, where the data elements are assigned to reducers according to their location in the file. Assuming locality, we expect most of the black circles to be close in space, and similarly for the white circles. Each reducer reports its MBRs, and then the β-clusters with overlapping MBRs are merged. The hope is that, due to locality, there will be much fewer pairs of overlapping β-clusters than in the random case, while enjoying even better load balancing.

5.5 Experimental Results

In this section, we describe the experiments performed to test the algorithms presented in the chapter. The experiments aimed at answering the following questions:

Q1 Among the reasonable choices described in Sect. 5.4, what is the best data partitioning approach?
Q2 How much (if at all) does the parallelism affect the clustering quality?
Q3 How does the parallel clustering method scale-up?
Q4 How accurate are the equations used in the cost-based optimization?

All experiments used the `Hadoop`[2] implementation for the `MapReduce` framework, on two `Hadoop` clusters: the M45 by Yahoo! and the DISC/Cloud by Parallel Data Lab in the Carnegie Mellon University. The M45 is one of the top 50 supercomputers in the world totaling 400 machines (3,200 cores), 1.5 PB of storage and 3.5 TB of main memory. The DISC/Cloud has 512 cores, distributed in 64 machines, 1TB of RAM and 256 TB of raw disk storage. The algorithm *Halite* was used as the plugged-in serial clustering method in all experiments.

---

[2] www.hadoop.com

Table 5.3 Summary of datasets

Dataset	Number of points	Number of axes	File size
YahooEig	1.4 billion	6	0.2 TB
TwitterEig	62 million	10	14 GB
Synthetic	up to 100 million	15	up to 14 GB

TB Terabytes, *GB* Gigabytes

The methods were tested over the real and synthetic datasets listed in Table 5.3, which are detailed as follows.

- YahooEig: The top 6 eigenvectors from the adjacency matrix of one of the largest web graphs. The web graph was crawled by Yahoo![3] in 2002 and contains 1.4 billion nodes and 6.6 billion edges. The eigenvectors amount to 0.2 TB.
- TwitterEig: The top 10 eigenvectors from the adjacency matrix of the Twitter[4] graph, that represents 62 million users and their relationships. The eigenvectors amount to 14 GB.
- Synthetic: A group of datasets with sizes varying from 100 thousand up to 100 million 15-dimensional points, containing 10 clusters each, and no noise. Clusters in subspaces of the original 15-dimensional space were created following standard procedures used by most of the clustering algorithms described in Chap. 3 , including the plugged-in serial clustering method used in the experiments. Specifically, Algorithm 6 was used again to generate the synthetic data. Axes-aligned clusters were created. Remember that the clusters generated by Algorithm 6 follow normal distributions with random mean and random standard deviation in at least 50% of the axes (relevant axes), spreading through at most 15% of the axes domains. In the other axes, the irrelevant ones, all clusters follow the uniform distribution, spreading through the whole axes domains.

Notice one remark: to evaluate how much (if at all) parallelism affects the serial clustering quality, the ideal strategy is to use as ground truth the clustering results obtained by running the plugged-in algorithm serially on any dataset, synthetic or real, and to compare these results to the ones obtained with parallel processing. However, for most of the large datasets analyzed in the experiments, to run a serial algorithm (*Halite* or, potentially, any other serial clustering method for moderate-to-high dimensionality data) is an impractical task—it would require impractical amounts of main memory and/or take a very long time. Thus, in practice, the Synthetic datasets are the only ones from which clustering ground truth is available, and they were used to evaluate the quality of all tested techniques in all experiments performed.

For a fair comparison with the plugged-in serial algorithm, the quality is computed following the same procedure used in Chap. 4. That is, the quality is computed by comparing the results provided by each technique to the ground truth, based on the averaged precision and recall of all clusters.

---

[3] www.yahoo.com

[4] http://twitter.com/

Table 5.4 Environmental parameters for M45

Parameter	Value
F_s	data file size
D_s	40 MB/s
N_s	20 MB/s
$start\_up\_cost(t)$	$0.1\,t$
$plug\_in\_cost(s)$	$1.4E^{-7}\,s$

The File-based Data Partitioning strategy may provide distinct quality results according to the order in which the input data is physically stored. Obviously, the best results appear when the data is totally ordered, i.e., the points of each cluster are sequentially stored in the data file. On the other hand, when the points are randomly distributed in the file, the qualities tend to be similar to those obtained by the approaches that use the Random-based Data Partitioning. For a fair analysis, each dataset from the Synthetic group was created considering an average case, i.e., 50% of the elements from the totally ordered case were randomly repositioned throughout the data file.

All experiments involving *BoW* were performed at M45. The parameters used are presented in Table 5.4. F_s refers to the data file size. D_s, N_s and $start\_up\_cost(t)$ were inferred by analyzing the logs of the M45 machines, while $plug\_in\_cost(s)$ was defined based on the time complexity analysis and experiments of the plugged-in method, which were previously presented in Chap. 4 .

The results on quality and wall-clock time reported for all experiments are the average of 10 *distinct runs*. A sample size of nearly one million elements (i.e., $S_r = \frac{1\ million}{\eta}$) was used in all experiments. Also, in every experiment the number of mappers m used was automatically defined by Hadoop.

5.5.1 Comparing the Data Partitioning Strategies

This section presents experiments that compare the data partitioning approaches. The experiments intend to answer question $Q1$: Among the reasonable choices in Sect. 5.4, what is the best data partitioning approach? In order to answer it, *ParC* was used, the most straightforward parallel algorithm described. This decision aims at avoiding that algorithmic characteristics influence the results, which should be related to the pros and to the cons of the data partitioning strategies only. That is, the intention is to avoid that the algorithm used leads to biased results.

Figures 5.5a and b show the quality of *ParC* using distinct data partitioning approaches (*ParC-F*—file-based, *ParC-R*–random-based and *ParC-L*–location-based) versus the wall-clock time over the Synthetic datasets, varying η and r respectively. The data sizes vary from 100 thousand to 100 million elements, while the number of reducers r starts at 2 and goes up to 16. The glyph sizes reflect the dataset size (Fig. 5.5a) or the number of reducers (Fig. 5.5b). Obviously, the ideal

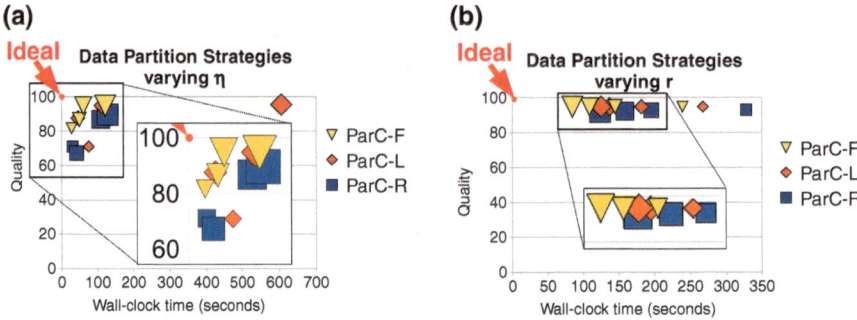

Fig. 5.5 File-based wins. Quality versus run time for *ParC* using distinct data partitioning approaches (*ParC-F*—file-based *yellow triangles*, *ParC-R*–random-based *blue squares* and *ParC-L*—location-based *orange diamonds*). *Left* 64 reducers, varying the data size $\eta = 100K$, 1M, 10M, 100M. *Right* 10 million elements dataset, varying the number of reducers $r = 2, 4, 8, 16$. The glyph sizes reflect the dataset size (**a**) or the number of reducers (**b**). *Top-left* is the ideal element—notice that *ParC-F* is consistently closer to it than the others. Thus, the file-based data partitioning approach is the one recommended to be used with *BoW*.

elements are in the top left of both plots, which represent 100 % quality obtained in zero time. Thus, the strategies were compared by analyzing how close they are to these elements in all cases. Notice that the three strategies present good quality, with some few exceptions for the Location and the Random-based ones. However, the File-based strategy consistently outperformed the others, presenting top quality and being the fastest one in all cases. The other Synthetic datasets generated very similar results. Thus, the *File-based Data Partition* is the partitioning strategy recommended to be used with *BoW*. The experiments presented in the rest of this chapter *always* employ this strategy.

We believe that the main reason for the success of the 'file-based' approach is that, when using this approach, the reducers process continuous pieces of the input file. This helps Hadoop to assign reduce tasks to the machines that already have the required data in local disks, turning the 'file-based' approach into the fastest one. Also, all experiments confirmed the hope that, due to locality, there will be much fewer pairs of overlapping β-clusters than in the random case, while enjoying even better load balancing.

5.5.2 Quality of Results

This section presents experiments that aim at answering question *Q2*: How much (if at all) does the parallelism affect the clustering quality? Figure 5.6 presents the quality results obtained by *ParC*, *SnI* and *BoW* over the Synthetic dataset with 10 million elements. All methods presented top quality, even for large numbers of reducers, like 1,024. Notice, that the serial execution quality of the plugged-in clustering method

Fig. 5.6 All variants give high quality results. 10 million dataset; quality versus number r of reducers for *ParC*, *SnI* and *BoW*. All methods match the quality of the serial clustering method (*top left*), for all values of r, like 1,024. The default, 'file-based' partitioning was used for all cases.

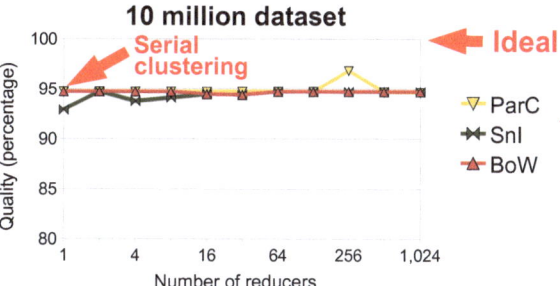

is the one obtained when using a single reducer ($r = 1$, extreme left elements in the plot). Similar results were observed with all Synthetic datasets.

An interesting observation is that the quality may decrease for small datasets, when using a large number of reducers. The obvious reason is that, in those cases, the method is partitioning a small amount of data through a large number of reducers, which actually receive too little data, not enough to represent the patterns existing in the dataset. This fact was confirmed in all experiments, and they lead to the recommendation of using at least $\sim$150 k points per reducer in average, that is, to set $r \leq \frac{\eta}{150\,k}$.

According to the experiments, the answer to question $Q2$ is: as long as you have enough data, the clustering quality is barely affected by the parallelism, even for extremely large numbers of reducers, such as, 1,024. *BoW* obtained top quality clusters in very little time from all of the very large datasets analyzed.

5.5.3 Scale-Up Results

This section presents experiments that aim at answering question $Q3$: How does the parallel clustering method scale-up? Scale-up results with different numbers of reducers are in Fig. 5.7. Here the TwitterEig eigenvectors and the Synthetic dataset with 100 million points were used. The plots show X-axes as the number of reducers r, and the Y-axes as the relative performance with n reducers compared to using 1 reducer (TwitterEig) or with 4 reducers (Synthetic). A fixed number of mappers $m = \sim 700$ was used. The results reported are the average of 10 distinct runs. 4 reducers were picked for the Synthetic dataset, as the running time using just one reducer was impractical. Note that the parallel clustering method exhibits the expected behavior: it starts with near-linear scale-up, and then flattens. Similar scale-up results were obtained for all other datasets.

The scale-up results with different data sizes are in Fig. 5.8. The YahooEig dataset is used. Random samples of the data with increasing sizes, up to the full dataset (1.4 billion elements) were generated to perform this experiment. Wall clock time versus data size is shown. The wall-clock time reported is the average time for 10

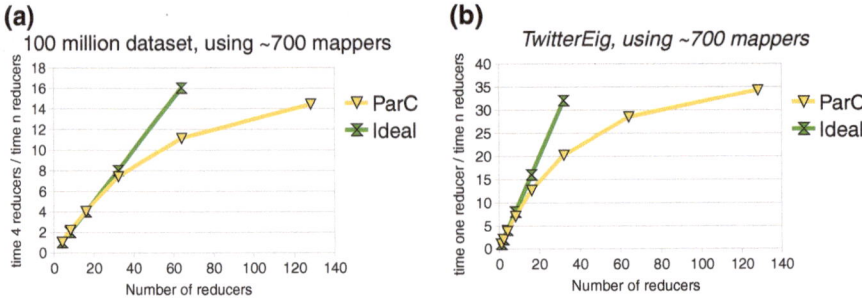

Fig. 5.7 Near-linear scale-up. Scale-up results regarding the number of reducers r. The parallel clustering method exhibits the expected behavior: it starts with near-linear scale-up, and then flattens. Numbers are the average of 10 runs for real and synthetic data. 100 million dataset (*left*); TwitterEig (*right*). The X-axes show the number of reducers r, and the Y-axes the relative performance with r reducers compared to using 1 reducer (*right*) or 4 reducers (*left*), in lin-lin scales. Using one reducer in the left case requires prohibitively long time. Number of mappers: $m = \sim 700$. The default, 'file-based' partitioning was used for all cases.

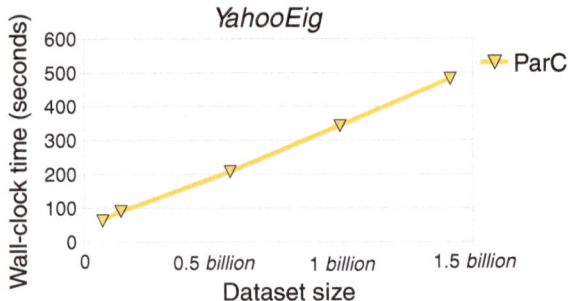

Fig. 5.8 Scale-up results: the parallel clustering method is linear on the dataset size. Wall-clock time (average of 10 runs) versus data size in lin-lin scales. Random samples from YahooEig, up to the full dataset (1.4 billion elements). Fixed number of reducers and mappers ($r = 128$ and $m = \sim 700$). The default, 'file-based' partitioning was used.

distinct runs. Fixed numbers of reducers and mappers ($r = 128$ and $m = \sim 700$) were used. As shown, the parallel clustering method has the expected scalability, scaling-up linearly with the data size.

It took only ~ 8 minutes to cluster the full dataset, which amounts to 200 GB. To provide some context to this result the time taken at different stages in the process is characterized: (a) the mappers took 47 seconds to read the data from disks; (b) 65 seconds were taken to shuffle the data; and (c) the reduce stage took 330 seconds. To estimate the time taken by the serial method in item (c), a random sample of the YahooEig dataset, of size $\frac{F_s}{r} = \frac{0.2TB}{128}$, was clustered by running the plug-in on a single machine (one core), similar to the ones of the used cloud of computes. The serial clustering time was 192 seconds. This indicates that the plug-in took $\sim 43\%$ of the total time. Similar results were obtained for all other datasets.

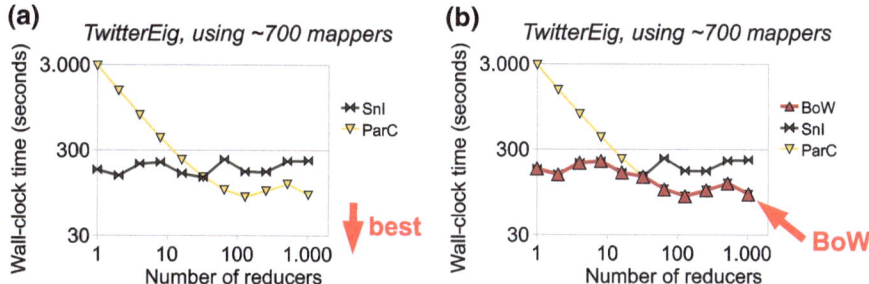

Fig. 5.9 *BoW* wins. Results for real data from Twitter. Wall-clock time versus number of reducers in log–log scale. ~700 `MapReduce` mappers were used for all runs. *Left: ParC* (*yellow down-triangles*) and *SnI* (*green butterflies*). The latter uses the novel *sampling-and-ignore* idea; *Right:* the same, *including* the method *BoW* (in *red up-triangles*). *BoW* achieves the best of both worlds, using cost-based optimization to pick the winning strategy and good parameters for it, and thus practically over-writes the corresponding curve on the graph.

5.5.4 Accuracy of the Cost Equations

Here we present experiments that illustrate the accuracy of the cost formulas shown in Eqs. 5.4 and 5.5 from Sect. 5.3, and the ability of *BoW* to choose the correct alternative. Figure 5.9 shows an example of *BoW*'s results on the `TwitterEig` dataset. It plots the wall-clock-time (average of 10 runs) versus the number of reducers, in log–log scales. Figure 5.9a shows the results for *ParC*, in yellow down-triangles, and *SnI*, in green 'butterfly' glyphs. The latter uses the novel *sampling-and-ignore* idea. Notice that there is no universal winner, with a cross-over point at about 30 machines for this setting. Figure 5.9b shows exactly the same results, this time *including* the wall-clock time of the method *BoW*, in red up-triangles. Notice that *BoW* locks onto the best of the two alternatives. The reason for its success is the cost-estimation formulas from Eqs. 5.4 and 5.5, which help *BoW* to pick the best alternative and to set good parameters for the chosen environment, while requiring nimble computational effort. Furthermore, notice that the two curves shown in *log–log scale* in Fig. 5.9a intersect at a narrow angle, which means that the optimal curve has a smooth plateau, and thus the cost is rather robust with respect to small variations of the environment parameters.

Figure 5.10 details the results for the Twitter data and also reports results for the `Synthetic` (100 million points) dataset, in the left and right columns respectively. The six plots give the wall-clock times (average of 10 runs) versus the number of reducers r, in log–log scales. The top row (a) and (d) shows that *BoW*, in red up-triangles, consistently picks the winning strategy among the two options: *ParC* (yellow down-triangles) and *SnI* (dark-green butterflies). For both datasets *BoW* gives results so close to the winner that its curve practically overwrites the winner's curve; the only overhead of *BoW* is the CPU time required to run the cost equations, which is obviously negligible.

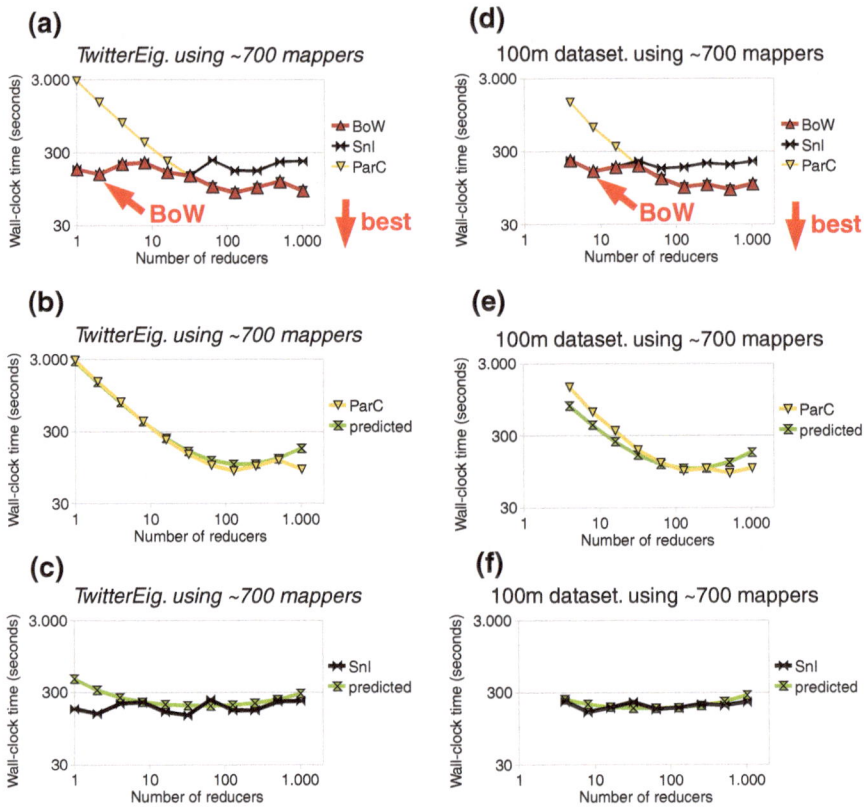

Fig. 5.10 *BoW* indeed achieves the *Best of both Worlds. BoW*'s results on the `TwitterEig` (*left*) and on the `Synthetic` 100 million (*right*) datasets. Time (average of 10 runs) versus number of reducers in log–log scale. *m* is ∼700 for all runs. *Top line*: illustration of *BoW*'s ability to pick the winner. Results for *ParC* (*yellow down-triangles*), for *SnI* (*green butterflies*) and for *BoW* (*red up-triangles*). Notice that *BoW* achieves the best of both worlds, consistently choosing the winning strategy, and thus practically over-writing the corresponding curve on those graphs. *Bottom two rows*: illustration of the accuracy of the Eqs. 5.4 and 5.5 for *ParC* and *SnI* respectively. In all cases, the green hour-glass shapes stand for the formulas; notice how close they are to the actual measurements (*yellow triangles*, and *dark-green butterfly shapes*, respectively)

The next two rows of Fig. 5.10 illustrate the accuracy of the cost formulas. Light-green hour-glasses indicate the theoretical prediction; yellow triangles stand for *ParC* in the middle row, and dark-green butterflies stand for *SnI* in the bottom row. Notice that the theory and the measurements usually agree very well. All other datasets provided similar results.

5.6 Conclusions

Given a *very large*, moderate-to-high dimensionality dataset, how could one cluster its points? For data that do not fit even on a single disk, parallelism is mandatory. In this chapter we described *BoW*, an algorithm that explores parallelism using MapReduce to cluster huge datasets. The main contributions of *BoW* are:

1. *Algorithm design and analysis*: the method *BoW* includes carefully derived cost functions that allow it to perform the automatic, dynamic trade-off between disk delay and network delay;
2. *Effectiveness, scalability and generality*: *BoW* has many desirable features. It can use almost any serial method as a plug-in (the only requirement: clusters described by hyper-rectangles), it uses no user defined parameters (due to its defaults), it matches the clustering quality of the serial algorithm, and it has near-linear scale-up;
3. *Experiments*: Experiments on both real and synthetic data including *billions* of points, and using up to 1,024 cores in parallel were performed. To the best of our knowledge, the *Yahoo! web* is the *largest real dataset ever reported* in the database clustering literature for moderate-to-high dimensionality data. *BoW* clustered its 200 GB in only 8 minutes, using 128 cores. Also, the experiments used up to 1,024 cores, which is again the highest such number published in the clustering literature for moderate-to-high dimensionality data.

This chapter presented one second algorithm that focuses on data mining in large sets of complex data. The next chapter describes in detail one third and last algorithm focused at tackling this hard problem.

References

1. Agrawal, R., Gehrke, J., Gunopulos, D., Raghavan, P.: Automatic subspace clustering of high dimensional data for data mining applications. SIGMOD Rec. 27(2), 94–105 (1998). http://doi.acm.org/10.1145/276305.276314
2. Agrawal, R., Gehrke, J., Gunopulos, D., Raghavan, P.: Automatic subspace clustering of high dimensional data. Data Min. Knowl. Discov. 11(1), 5–33 (2005). http://dx.doi.org/10.1007/s10618-005-1396-1
3. Cordeiro, R.L.F., Traina, A.J.M., Faloutsos, C., Traina Jr., C.: Finding clusters in subspaces of very large, multi-dimensional datasets. In: F. Li, M.M. Moro, S. Ghandeharizadeh, J.R. Haritsa, G. Weikum, M.J. Carey, F. Casati, E.Y. Chang, I. Manolescu, S. Mehrotra, U. Dayal, V.J. Tsotras (eds.) ICDE, pp. 625–636. IEEE (2010).
4. Cordeiro, R.L.F., Traina, A.J.M., Faloutsos, C., Traina Jr., C.: Halite: Fast and scalable multi-resolution local-correlation clustering. IEEE Transactions on Knowledge and Data Engineering 99(PrePrints) (2011). http://doi.ieeecomputersociety.org/10.1109/TKDE.2011.176. 16 pages
5. Cordeiro, R.L.F., Traina Jr., C., Traina, A.J.M., López, J., Kang, U., Faloutsos, C.: Clustering very large multi-dimensional datasets with mapreduce. In: C. Apté, J. Ghosh, P. Smyth (eds.) KDD, pp. 690–698. ACM (2011).
6. Fayyad, U.: A data miner's story - getting to know the grand challenges. In: Invited Innovation Talk, KDD (2007). Slide 61. Available at: http://videolectures.net/kdd07_fayyad_dms/

7. Moise, G., Sander, J.: Finding non-redundant, statistically significant regions in high dimensional data: a novel approach to projected and subspace clustering. In: KDD, pp. 533–541 (2008).

8. Moise, G., Sander, J., Ester, M.: P3C: A robust projected clustering algorithm. In: ICDM, pp. 414–425. IEEE Computer Society (2006).

9. Moise, G., Sander, J., Ester, M.: Robust projected clustering. Knowl. Inf. Syst. 14(3), 273–298 (2008). http://dx.doi.org/10.1007/s10115-007-0090-6

10. Yiu, M.L., Mamoulis, N.: Iterative projected clustering by subspace mining. TKDE **17**(2), 176–189 (2005). doi:10.1109/TKDE.2005.29

Chapter 6
QMAS

Abstract This chapter describes a work that uses the background knowledge of the clustering algorithms previously presented in the book to focus on two *distinct* data mining tasks-the tasks of labeling and summarizing large sets of complex data. Given a large collection of complex objects, *very few* of which have labels, how can we guess the labels of the remaining majority, and how can we spot those objects that may need brand new labels, different from the existing ones? The work presented here provides answers to these questions. Specifically, this chapter describes in detail *QMAS* [2], one third algorithm that focuses on data mining in large sets of complex data, which is a fast and scalable solution to the problem of automatically analyzing, labeling and understanding this kind of data.

Keywords Complex data · Low-labor labeling · Summarization · Attention routing · Correlation clustering · Random walks with restarts · Satellite imagery analysis

6.1 Introduction

The problem of automatically analyzing, labeling and understanding large collections of complex objects appears in numerous fields. One example application refers to satellite imagery, involving a scenario in which a topographer wants to analyze the terrains in a collection of satellite images. Let us assume that each image is divided into tiles (say, 16×16 pixels). Such a user would like to label a small number of tiles ('Water', 'Concrete', etc), and then the ideal system would automatically find labels for all the rest. The user would also like to know what strange pieces of land exist in the analyzed regions, since they may indicate anomalies (e.g., de-forested areas, potential environmental hazards, etc.), or errors in the data collection. Finally, the user would like to have a few tiles that best represent each kind of terrain.

R. L. F. Cordeiro et al., *Data Mining in Large Sets of Complex Data*,   93
SpringerBriefs in Computer Science, DOI: 10.1007/978-1-4471-4890-6_6,
© The Author(s) 2013

Fig. 6.1 One example satellite image of Annapolis (MD, USA), divided into 1,024 (32 × 32) tiles, only 4 of which are labeled with keywords, "City" (*red*), "Water" (*cyan*), "Urban trees" (*green*) or "Forest" (*black*). IKONOS/GeoEye-1 Satellite image courtesy of GeoEye

Figure 6.1 illustrates the problem on the example application of satellite images. It shows an example image from the city of Annapolis, MD, USA, decomposed into 1,024 (32 × 32) tiles, very few (only four) of which were manually labeled as "City" (red), "Water" (cyan), "Urban Trees" (green) or "Forest" (black). With this input set, we want a system that is able to automatically assign the most appropriate labels to the unlabeled tiles, and provide a summarized description of the data by finding clusters of tiles, the N_R best representatives for the data patterns and the top-N_O outlier tiles.

Similar requirements appear in several other settings that involve distinct types of complex data, such as, social networks, and medical image or biological image applications. In a social network, one user wants to find other users that share similar interests with himself/herself or with his/her contacts, while the network administrator wants to spot a few example users that best represent both the most typical and the most strange types of users. In medicine, physicians want to find tomographies or x-rays similar to the images of their patient's as well as a few examples that best represent both the most typical and the most strange image patterns. In biology, given a collection of fly embryos [10] or protein localization patterns [7] or cat retina images [1] and their labels, biologists want a system to answer similar questions.

The goals of this chapter are summarized in two research problems:

Problem 6.1 *low-labor labeling (LLL)* - **Given** an input set $I = \{I_1, I_2, I_3, \ldots, I_{N_I}\}$ of N_I complex objects, *very few* of which are labeled with keywords, **find** the most appropriate labels for the remaining ones.

Problem 6.2 *mining and attention routing* - **Given** an input set $I = \{I_1, I_2, I_3, \ldots, I_{N_I}\}$ of N_I partially labeled complex objects, **find** clusters, the N_R objects from I that best represent the data patterns and the top-N_O outlier objects.

This chapter describes in detail *QMAS*: *Q*uerying, *M*ining *A*nd *S*ummarizing *Multi-dimensional Databases* [2]. The method is a fast ($O(N)$) solution to the aforementioned problems. Its main contributions, supported by experiments on real satellite images, spanning up to more than 2.25 GB, are summarized as follows:

1. **Speed**: *QMAS* is fast and it scales linearly on the database size, being up to 40 times faster than top related works on the same subject;
2. **Quality**: It can do *low-labor labeling (LLL)*, providing results with better or equal quality when compared to top related works;
3. **Non-labor intensive**: It works even when it is given very few labels - *it can still extrapolate from tiny sets of pre-labeled data*;
4. **Functionality**: Contrasting to other methods, *QMAS* encompasses extra mining tasks such as clustering and outlier and representatives detection as well as summarization. It also spots data objects that potentially require new labels;

6.2 Presented Method

This section describes *QMAS*, one fast and scalable solution to the problems of *low-labor labeling (LLL)* (Problem 6.1) and *mining and attention routing* (Problem 6.2). It assumes that feature extraction is first applied over the input set of complex objects I, turning the set into a multi-dimensional dataset. Next, we detail the method.

6.2.1 Mining and Attention Routing

In this section we describe the solution to the problem of *mining and attention routing* (Problem 6.2). The general idea is as follows: First, *QMAS* does clustering on the input set of complex objects I; then, it finds (a) the subset of objects $R = \{R_1, R_2, R_3, \ldots, R_{N_R}\} \mid R \subseteq I$ that best represent I, and (b) the array with the top-N_O outlier objects $O = (O_1, O_2, O_3, \ldots, O_{N_O}) \mid O_o \in I, \forall\ 1 \le o \le N_O$, sorted according to the confidence degree of each object O_o be an outlier. Algorithm 10 provides a general view of the solution to Problem 6.2. The details are given as follows.

Algorithm 10 : *QMAS-mining.*

Input: input set of complex objects I; number of representatives N_R; number of top outliers N_O.
Output: clustering result C; set of representatives $R = \{R_1, R_2, R_3, \ldots, R_{N_R}\} \mid R \subseteq I$;
        top-N_O outliers $O = (O_1, O_2, O_3, \ldots, O_{N_O}) \mid O_o \in I, \forall\, 1 \leq o \leq N_O$, in sorted order.
1: do clustering on I, let the result be C;
2: R = random N_R complex objects from I;
3: *error* $= E_{QMAS}(I, R)$; // from Equation 6.2
4: **repeat**
5:    improve the representatives in R;
6:    *old_error* = *error*;
7:    *error* $= E_{QMAS}(I, R)$; // from Equation 6.2
8: **until** *error* == *old_error*
9: $O =$ the N_O complex objects from I worst represented by R, sorted according to the confidence
   degree of each object O_o in O be an outlier;
10: **return** C, R and O;

6.2.1.1 Clustering

The clustering step over the input set of objects I is performed using the algorithm *Halite* [3, 4]. As described in Chap. 4, *Halite* is a fast and scalable clustering algorithm well-suited to spot clusters in large collections of complex data. *QMAS* takes advantage of the soft clustering process of Halite to allow a single object to belong to one or more clusters with equal probabilities.

This configuration of Halite is used by QMAS to find clusters in the objects of I.

6.2.1.2 Finding Representatives

Now we focus on the problem of selecting a set $R = \{R_1, R_2, R_3, \ldots, R_{N_R}\} \mid R \subseteq I$ of objects with cardinality $N_R = |R|$ to represent the input set of complex objects I. First, we discuss the desirable properties for a set of representatives, then we describe two possible approaches to actually find the representatives.

An appropriate set of representatives R for the objects in I must have the following property: *there is a large similarity between every object $I_i \in I$ and its most similar representative R_r*. Obviously, the set of representatives that best represent I is the full set of complex objects, $N_R = N_I \Rightarrow R = I$. In this case, the similarity is maximal between each object $I_i \in I$ and its most similar representative R_r, which is the complex object itself, $I_i = R_r$. However, for $N_R < N_I$, how should one evaluate the quality of a given set of representatives?

A simple way to evaluate the quality of a collection of representatives is to sum the squared distances between each object I_i and its closest representative R_r. This gives us an error function that should be minimized in order to achieve the best set of representatives R for the input set of complex objects I. It is not a coincidence that this is the same error function minimized by the classic clustering algorithm K-Means [8, 9, 11], which is formally defined by the following equation.

$$E_{KM}(I, R) = \sum_{I_i \in I} MIN\{\|I_i - R_r\|^2 \mid R_r \in R\} \qquad (6.1)$$

In the equation, $\|I_i - R_r\|$ is the distance between the objects I_i and R_r, and MIN is a function that returns the minimum value within its input set of values. Without loss of generality, the Euclidean distance L_2 is considered here.

Based on this idea, when we ask K-Means for N_R clusters, the centroids of the clusters are good indicators of the data space positions where we should look for representatives. Then, we have a set of representatives for K-Means by: (1) finding, for each centroid, the data object I_i from I that is the closest one to the respective centroid, and; (2) defining R to be the complete set of objects found.

Figure 6.2a shows a synthetic dataset containing three clusters. The clusters and their sizes follow skewed distributions. The sizes are 30,000, 3,000 and 1,000 for the clusters in the bottom left, bottom right and top of the data space respectively. 500 points are uniformly distributed through the data space to represent noise.

Figure 6.2b shows the representatives selected for the toy dataset by using K-Means and considering N_R as 10 (top) and 20 (bottom). The results presented are the best ones over 50 runs, i.e., the ones with the smallest error computed by Eq. 6.1. Notice that, in all cases, the representatives selected are excessively concentrated in the bottom left cluster, the biggest one, while the other two clusters are poorly represented, having only a few representatives each. These results indicate that K-Means is sensitive to the data distribution, commonly presenting unsatisfactory results for representative picking, especially for skewed data distributions.

$QMAS$ proposes to *use* the traditional, well-known K-Harmonic Means clustering algorithm [12] for representative picking, since it is almost insensitive to skewed

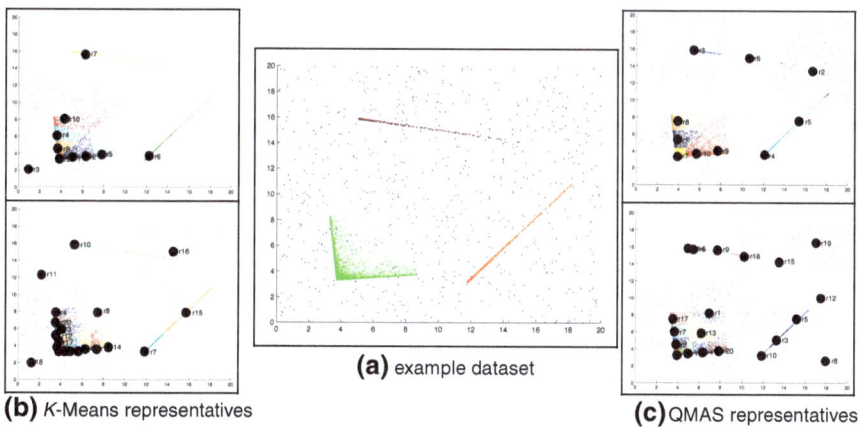

(a) example dataset

(b) K-Means representatives   **(c)** QMAS representatives

Fig. 6.2 Examples of representatives in synthetic data. *Center*: toy dataset with 3 clusters following skewed distributions; *borders*: representatives selected by K-Means (*left*) and $QMAS$ (*right*), for $N_R = 10$ (*top*) and 20 (*bottom*). These are the results with the smallest error over 50 runs

distributions, data imbalance, and bad seed initialization. Thus, it is a robust way to look for representatives, again by asking for N_R clusters and, for each cluster, picking as a representative the object I_i from I that is the closest object to the respective cluster's centroid. The minimization error function is presented as follows.

$$E_{QMAS}(I, R) = \sum_{I_i \in I} HAR\{\|I_i - R_r\|^2 \mid R_r \in R\} = \sum_{I_i \in I} \frac{N_R}{\displaystyle\sum_{R_r \in R} \frac{1}{\|I_i - R_r\|^2}} \quad (6.2)$$

In the equation, $\|I_i - R_r\|$ is the distance between the data objects I_i and R_r, and HAR is a function that returns the harmonic mean of its input values. The Euclidean distance L_2 is used once more, without loss of generality.

Figure 6.2c shows the representatives selected by $QMAS$ for the toy dataset, again considering N_R as 10 (top) and 20 (bottom). Once more, the results presented are the best ones over 50 runs, this time considering the error function in Eq. 6.2. Notice that the representatives chosen are now well distributed among the three clusters, providing to the user a summary that better describes the data patterns.

6.2.1.3 Finding the Top-N_O Outliers

The final task related to the problem of *mining and attention routing* is to find the top-N_O outliers $O = (O_1, O_2, O_3, \ldots, O_{N_O}) \mid O_o \in I, \forall\, 1 \leq o \leq N_O$, for the input set of complex objects I. In other words, O contains the N_O objects of I that diverge the most from the main data patterns. The outliers must be sorted in such a way that we identify the top $1st$ outlier, the top $2nd$ outlier and so on, according to the confidence degree of each one being an outlier.

To achieve this goal, $QMAS$ takes the representatives found in the previous section as a base for the outliers definition. Assuming that a set of representatives R is a good summary for I, the N_O objects from I that are the worst represented by R are said to be the top-N_O outliers. Consider again the error function in Eq. 6.2. Notice that the minimized error is the summation of the individual errors for each object $I_i \in I$, where the individual error for I_i is given by the following equation.

$$IE_{QMAS}(I_i, R) = \frac{N_R}{\displaystyle\sum_{R_r \in R} \frac{1}{\|I_i - R_r\|^2}} \quad (6.3)$$

This equation is the harmonic mean of the squared distances between one object I_i and each one of the representative objects in R. The object $I_i \in I$ with the greatest individual error is the one that is worst represented by R, which is the object considered to be the top $1st$ outlier of I. The top $2nd$ outlier is the object with the

Fig. 6.3 Top-10 outliers for the toy dataset in Fig. 6.2a, using the QMAS' representatives from Fig. 6.2c (*top*). As we can see, the *top* outliers are actually the most extreme cases for this data

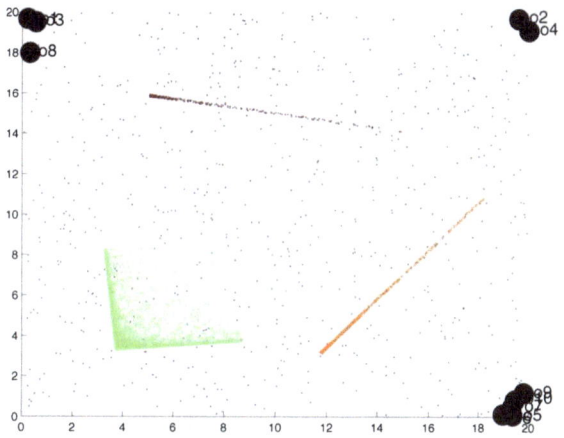

second greatest individual error, and so on. In this way, *QMAS* defines the array O containing the top-N_O outliers, in sorted order.

Figure 6.3 shows the top-10 outliers that *QMAS* found for the example dataset in Fig. 6.2a, considering $N_O = 10$ and $N_R = 10$. As we can see, the top outliers are actually the most extreme cases for this data.

6.2.2 Low-Labor Labeling (LLL)

In this section we discuss the solution to the problem of *low-labor labeling (LLL)* (Problem 6.1). That is, given the input set $I = \{I_1, I_2, I_3, \ldots, I_{N_I}\}$ of N_I complex objects, very few of which are labeled with keywords, how to find the most appropriate labels for the remaining ones. In order to tackle this problem, *QMAS* first represents the input complex objects and labels as a graph G, which is named as the *Knowledge Graph*. Then, random walks with restarts over G allow *QMAS* to find the most suitable labels for each unlabeled object. Algorithm 11 provides a general view of the solution to Problem 6.1. The details are given as follows.

G is a tri-partite graph composed of a set of vertexes V and a set of edges X, i.e., $G = (V, X)$. To build the graph, the input sets of complex objects I and known labels L are used, as well as the clustering results obtained in Sect. 6.2.1. V consists of one vertex for each data object, for each cluster, and for each label. The edges link complex objects to their respective clusters and labels. Let $V(I_i)$ and $V(L_l)$ represent the vertexes of G related to object I_i and to label L_l respectively. Provided the clustering results for the objects in I, the process of building G is simple, having linear time and memory complexities on the number of objects, labels and clusters.

Figure 6.4 shows the *Knowledge Graph G* for a small example dataset with seven complex objects, two labels, and three clusters. Data objects, labels, and clusters

---

Algorithm 11 : *QMAS-labeling.*

Input: input collection of complex objects I; collection of known labels L;
     restart probability c; clustering result C. // from Algorithm 10
Output: full set of labels LF.
1: use I, L and C to build the *Knowledge Graph* G;
2: **for** each unlabeled object $I_i \in I$ **do**
3:    do random walks with restarts in G, using c, always restarting the walk from vertex $V(I_i)$;
4:    compute the affinity between each label in the collection L and the object I_i. Let L_l be the
      label with the biggest affinity to I_i;
5:    set in LF: L_l is the appropriate label for object I_i;
6: **end for**
7: **return** LF;

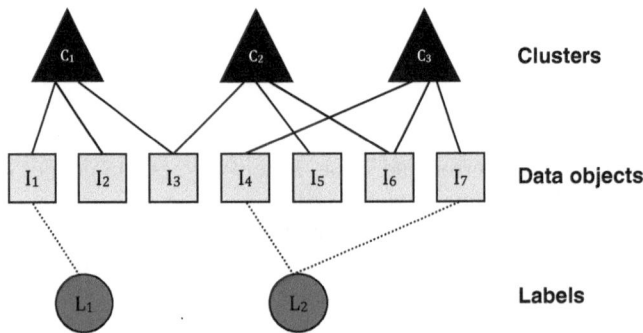

Fig. 6.4 The *knowledge graph G* for a toy dataset. *Squares, circles,* and *triangles* represent data objects, labels, and clusters respectively. The edges link objects to their corresponding clusters and known labels

are represented by nodes with shape of squares, circles, and triangles, respectively. The graph indicates, for example, that cluster C_1 contains the objects I_1, I_2, and I_3. Object I_3 also belongs to cluster C_2 in this setting. In addition, the graph shows that object I_1 has the known label L_1, while the objects I_4 and I_7 have the known label L_2.

In order to look for the most suitable label for an unlabeled complex object I_i, QMAS uses random walks with restarts over graph G. This process works as follows: a random walker starts from vertex $V(I_i)$. At each step, the walker either goes back to the initial vertex $V(I_i)$, with probability c, or to a randomly chosen vertex that shares an edge with the current vertex, with probability $1 - c$. The value of c is user defined, and may be determined by cross validation. The probability of choosing a neighboring vertex is proportional to the degree of that vertex, *i.e.*, the walker favors smaller clusters and more specific labels in this process. The affinity between I_i and a label L_l is given by the steady state probability that the random walker will find itself at vertex $V(L_l)$, always restarting from $V(I_i)$. Finally, the label L_l with the largest affinity with object I_i is taken as the most suitable label for I_i.

The intuition behind this procedure is that the steady state probability that a random walker will find itself in vertex $V(L_l)$, always restarting the walk from vertex $V(I_i)$, is a way to measure the closeness between I_i and L_l. If the computed probability is high, the vertexes are probably linked by short paths. On the other hand, if the probability is low, it is likely that no short path links them. This idea can be better understood through the example in Fig. 6.4. Let us assume that we want to find the most appropriate label for object I_2. There is a high probability that a random walker will reach $V(L_1)$, always restarting the walk from $V(I_2)$, mainly because there exists a three-step path linking $V(I_2)$ to $V(L_1)$. On the other hand, there is a lower probability that the walker will find itself at $V(L_2)$, always restarting the walk from $V(I_2)$, especially because the shortest path between $V(I_2)$ and $V(L_2)$ has seven steps. It leads us to conclude that the most appropriate label for I_2 is L_1.

6.3 Experimental Results

This section discusses the experiments performed to test the *QMAS* algorithm. Large collections of satellite images were analyzed to validate the method. First, results on the initial example from the introductory Sect. 6.1 are reported. Then, we discuss the experiments performed to support the contributions of *QMAS* stated in that section, regarding its Speed, Quality, Non-labor intensive capability, and Functionality.

Three real satellite image sets were analyzed. They are described as follows:

- *GeoEye*[1]—this public dataset contains 14 high-quality satellite images in the jpeg format extracted from famous cities around the world, such as the city of Annapolis (MD, USA), illustrated in Fig. 6.1. The total data size is about 17 MB. Each image was divided into equal-sized rectangular tiles and the entire dataset contains 14,336 tiles, from which Haar wavelets features in 2 resolution levels were extracted, plus the mean value of each band of the tiles;
- *SAT1.5 GB*—this proprietary dataset has 3 large satellite images of around 500 MB each in the GeoTIFF lossless data format. The total data size is about 1.5 GB. Each image was divided into equal-sized rectangular tiles. The 3 images combined form a set of 721,408 tiles, from which Haar wavelets features in 2 resolution levels were extracted, plus the mean value of each band of the tiles;
- *SATLARGE*—this proprietary dataset contains a pan QuickBird image of size 1.8 GB, and its matching 4-band multispectral image of size 450 MB. These images were combined and 2,570,055 hexagonal tiles generated, from which mean, variance, moments and GBT texture features [5] were extracted. The final feature set of a tile comprises a 30-dimensional vector.

The experimental environment is a server with Fedora® Core 7 (Red Hat, Inc.), a 2.8 GHz core and 4 GB of RAM. *QMAS* was compared with one of the best competitors: the GCap method that we described in Sect. 2.4. GCap was implemented in two versions with different nearest neighbor finding algorithms: one version uses the basic quadratic algorithm (GCap) and one other version spots approximate nearest

---

[1] The data is publicly available at: 'geoeye.com'.

neighbors (GCap-ANN), using the ANN Library.[2] The number of nearest neighbors is set to seven in all experiments. All three approaches share the same implementation of the Random Walks with Restarts algorithm using the *power iteration method* [6], with the restart parameter set as $c = 0.15$.

6.3.1 Results on the Initial Example

This section discusses the results obtained for the example satellite image from Fig. 6.1, presented in the introductory Sect. 6.1. The image, also shown in Fig. 6.5a, refers to the city of Annapolis, MD, USA. As in the introductory example, it was decomposed into 1,024 (32×32) tiles, *only four* of which were manually labeled as "City" (red), "Water" (cyan), "Urban Trees" (green) or "Forest" (black). From each tile Haar wavelets features in 2 resolution levels were extracted, plus the mean value of each band of the tile.

Figure 6.5b shows the solution proposed by *QMAS* to the problem of *low-labor labeling (LLL)* (Problem 6.1) on the example satellite image. Notice two remarks: (a) the vast majority of the tiles were correctly labeled and (b) there are few outlier tiles marked in yellow that *QMAS* judges as too different from the labeled ones (i.e., there is no path linking the image and one label in the *Knowledge Graph*), and thus are returned to the user as outliers that potentially deserve a new label of their own. Closer inspection shows that the outlier tiles tend to be on the border of, say, "Water" and "City" (because they contain a bridge).

The solution to the problem of *mining and attention routing* (Problem 6.2) on the example image is presented in Fig. 6.5c and d. *QMAS* pointed out the 3 tiles that best represent the data patterns and the top-2 outliers. Notice that the representatives actually cover the 3 major keywords ("City", "Urban Trees", and "Water"), while the top outliers are hybrid tiles, like the bottom right which is a bridge (both "Water" and "City").

Note that *QMAS* goes even further by summarizing the results: besides representatives and top outliers, *QMAS* found data clusters, ignoring the user-provided labels. This has two advantages. The first is that it indicates to the user what, if any, changes have to be done to the labels: new labels may need to be created (to handle some clusters or outliers), and/or labels may need to be merged (e.g., "Forest" and "Urban trees"), and/or labels that are too general may need to be divided in two or more ("Shallow Water" and "Deep Sea", instead of just "Water"). The second advantage is that these results can also be used for group labeling, since the user can decide to assign labels to entire clusters rather than labeling tiles one at a time.

---

[2] http://www.cs.umd.edu/~mount/ANN/

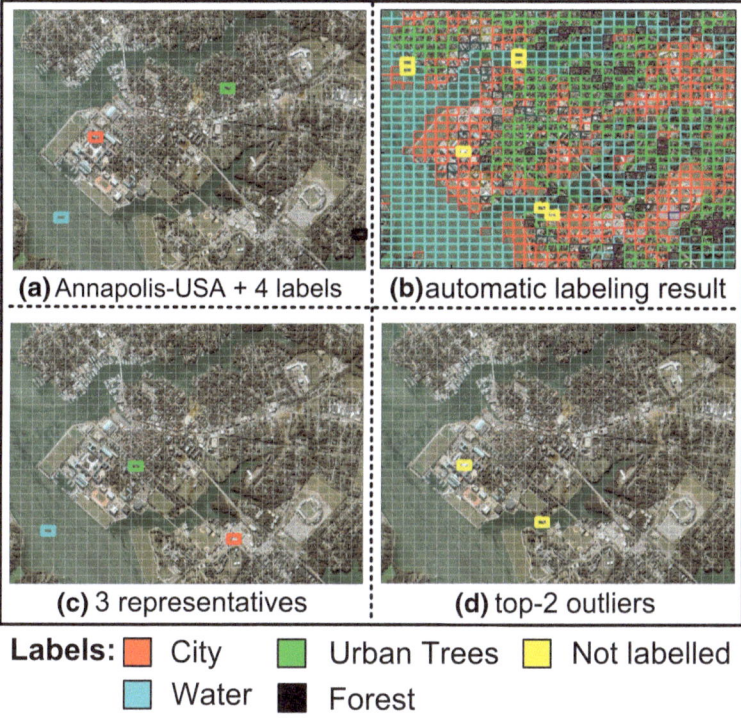

Labels: ☐ City ☐ Urban Trees ☐ Not labelled
☐ Water ■ Forest

Fig. 6.5 Solution to the problems of *low-labor labeling* and *mining and attention routing* on an example satellite image. *Top left*: the input image of Annapolis (MD, USA), divided into 1,024 (32 × 32) tiles, only 4 of which are labeled with keywords ("City" in red, etc). *Top right*: the labels that *QMAS* proposes; *yellow* indicates outliers. *Bottom left*: the 3 tiles that best represent the data, which actually cover the 3 major keywords. *Bottom right*: the top-2 outlier tiles, where appropriate labels do not exist (hybrid tiles, like the *bottom right* which is a bridge = both "Water" and "City"). IKONOS/GeoEye-1 Satellite image courtesy of GeoEye

6.3.2 Speed

This section discusses experiments that support the following claim: *QMAS* is a fast solution to the problems investigated, scaling linearly on the data size, and being several times faster than top related works. Figure 6.6 shows how the tested methods scale with increasing data sizes. Random samples from the *SAT1.5* GB dataset were used. As it can be seen, the log-log curve for QMAS has the slope equal to one, so *QMAS* scales linearly with the input data size, while the slope of log-log curves are 2.1 and 1.5 for GCap and GCap-ANN, respectively. For the full *SAT1.5* GB dataset, *QMAS* is 40 times faster than GCap-ANN, while running GCap would take hours long (not shown in the figure).

Fig. 6.6 Time versus number of tiles for random samples of the *SAT1.5* GB dataset. *QMAS*: *red circles*; GCap: *blue crosses*; GCap-ANN: *green diamonds*. Wall-clock time results are averaged over 10 runs; Error bars are too small to be shown

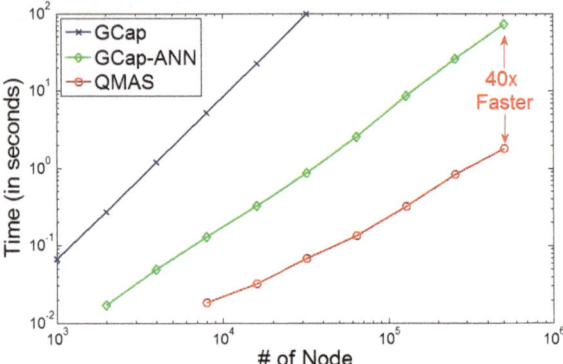

Notice one *important* remark: as stated in Sect. 2.4, most previous works, including GCap, search for nearest neighbors in the feature space. This operation is super-linear even with the use of approximate nearest-neighbor finding algorithms. On the other hand, *QMAS* avoids the nearest neighbor searches by using clusters to link similar image nodes in the *Knowledge Graph*. It allows *QMAS* to scale linearly on the data size, being up to 40 times faster than the top competitors.

6.3.3 Quality and Non-labor Intensive

This section discusses experiments that support the ability of *QMAS* to return high-quality results and to be non-labor intensive. For the experiments, 256 tiles in the *SAT1.5* GB dataset were labeled via manual curation. Some few ground truth labels were randomly selected from each class as the input labels and the remaining ones were used for one quality test. Figure 6.7 illustrates the labeling accuracy for the GCap-ANN and for the *QMAS* approaches in box plots obtained from 10 repetitive runs. As it can be seen, *QMAS* does not sacrifice quality for speed compared with GCap-ANN and it performs even better when the pre-labeled data size is limited. Note that the accuracy of *QMAS* is barely affected by the number of the pre-labeled examples in each label class, when the number of examples given goes above 2, while the quality of GCap-ANN was considerably worse with small sets of pre-labeled examples. The fact that *QMAS* can still extrapolate from tiny sets of pre-labeled data ensures its non-labor intensive capability.

6.3.4 Functionality

This section discusses experiments that support the following claim: in contrast to the related works, *QMAS* includes other mining tasks such as clustering, detection

Fig. 6.7 Comparison of approaches in box plots— quality versus size of the pre-labeled data. *Top left* is the ideal point. *QMAS: red circles*; GCap-ANN: *green diamonds*. Accuracy values of *QMAS* are barely affected by the size of the pre-labeled data. Results are obtained over 10 runs

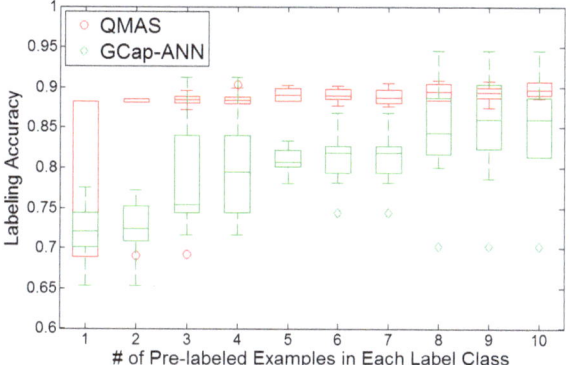

of top outliers and representatives, besides summarization. In other words, *QMAS* tackles both the problem of *low-labor labeling* (*LLL*) (Problem 6.1) and the problem of *mining and attention routing* (Problem 6.2), while the related works address only the former. To evaluate this claim, the functionality of *QMAS* was analyzed regarding its ability to spot clusters, representatives and top outliers. The *GeoEye* dataset was used in all experiments of this section.

Figure 6.8 shows some screenshots of the clustering results obtained with *QMAS*. Yellow tiles represent outliers. Closer inspection shows that these outlier tiles tend to be on the border of areas like "Water" and "City" (because they contain a bridge). The remaining tiles are colored according to its cluster. As expected, a few tiles belong to more than one cluster, since *QMAS* does soft clustering. These tiles were colored

Fig. 6.8 Clustering results provided by *QMAS* for the *GeoEye* dataset. *Top*: the real satellite images; *bottom*: the corresponding results, shown by coloring each tile after its cluster. *Yellow tiles* represent outliers. Notice that the clusters actually represent the main data patterns. IKONOS/GeoEye-1 Satellite image courtesy of GeoEye

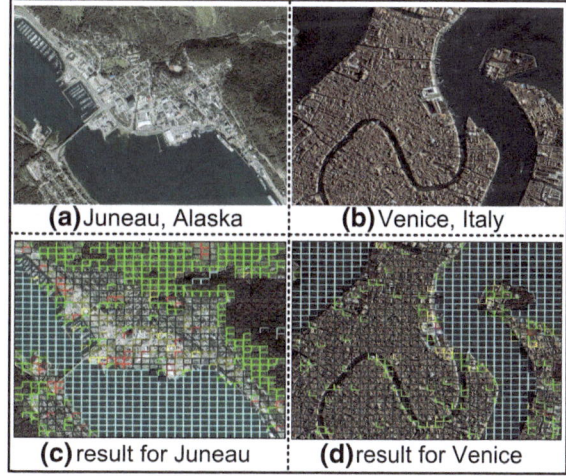

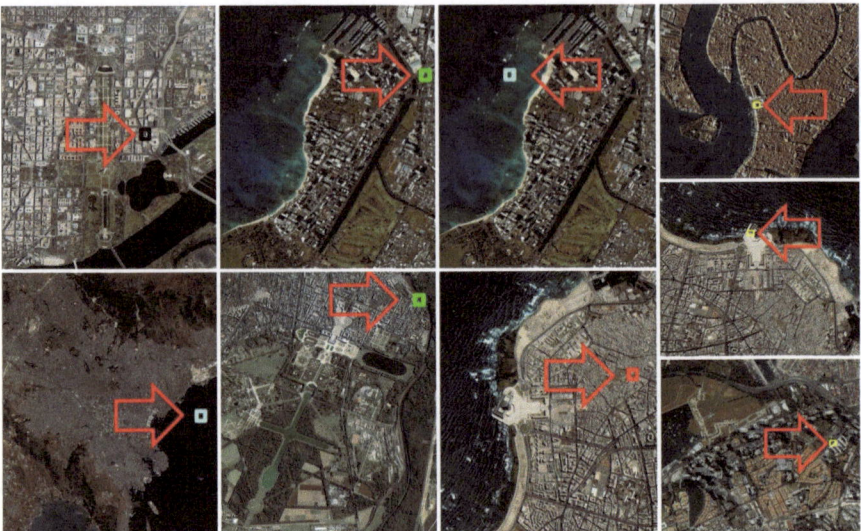

Fig. 6.9 *Left*: $N_R = 6$ representatives found by *QMAS* for the *GeoEye* dataset, colored after their clusters. *Right*: Top-3 outliers for the same dataset, found using the representatives shown. Notice that the 3 outliers together with the 6 representatives found, only 9 tiles in total, nicely summarize the *GeoEye* dataset, which contains more than 14 thousand tiles. IKONOS/GeoEye-1 Satellite image courtesy of GeoEye

according to their first assigned clusters. Notice: the clustering results reported indeed represent the main patterns apparent in the analyzed images.

Figure 6.9 illustrates the results obtained by *QMAS* for representatives (left) and top outliers (right). $N_R = 6$ representatives are shown, colored according to their clusters. Note that these few representatives cover the main clusters previously presented in Fig. 6.8. Also, these 6 representatives were used as a basis to the detection of the top-3 outliers shown in the figure. The outlier tiles tend to be on the border of areas like "Water" and "City" (because they contain a bridge). By comparing these results with the clusters shown in Fig. 6.8, one can easily notice that the 3 outliers spotted, together with the 6 representatives found (only 9 tiles in total) properly summarize the *GeoEye* dataset, which has more than 14, 000 tiles.

6.3.5 Experiments on the *SATLARGE* Dataset

Here we discuss results for the *SATLARGE* dataset, related to *query by example* experiments; *i.e.*, given a small set of tiles (examples), manually labeled with one keyword, query the unlabeled tiles to find the ones most likely related to that keyword. Figure 6.10 illustrates the results obtained for several categories ("Water", "Houses", "Trees", "Docks", "Boats" and "Roads") to show that *QMAS* returns high-quality

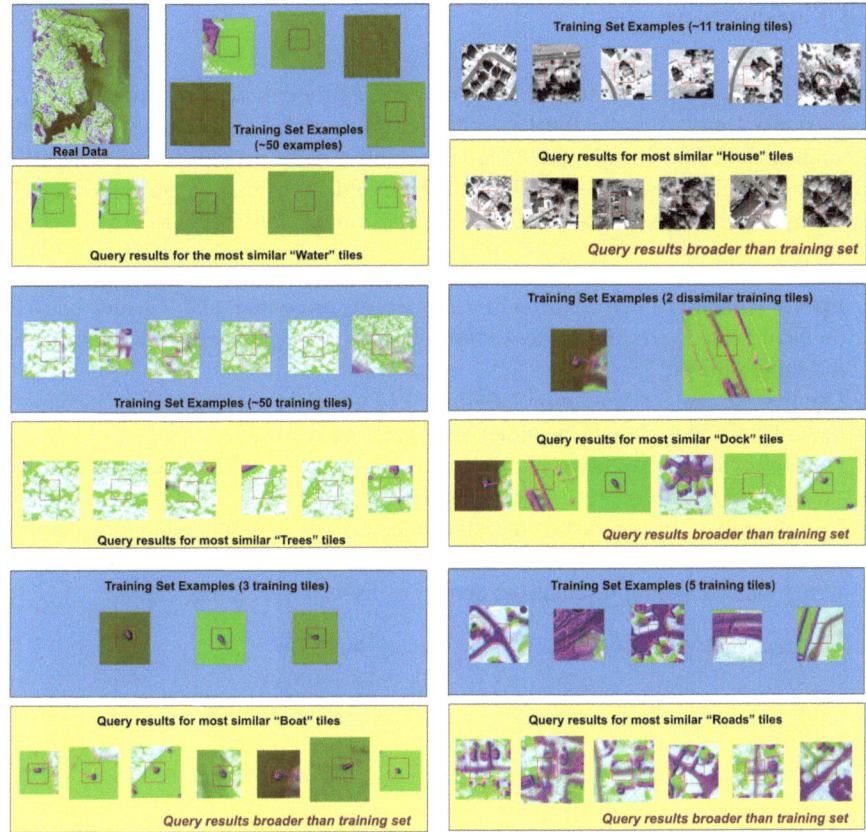

Fig. 6.10 Examples with "Water", "Houses", "Trees", "Docks", "Boats" and "Roads": labeled data and the results of queries aimed at spotting other tiles of these types. IKONOS/GeoEye-1 Satellite image courtesy of GeoEye

results, being almost insensitive to the kind of tile given as input. Also, notice in the experiments related to "Docks" and "Boats" that the results are correct even for tiny sets of pre-labeled data. The number of examples provided vary from as many as ~50 examples to as few as *two* examples. Varying the amount of labeled data allows one to observe how the system responds to these changes. In general, labeling only a small number of examples (even less than five) still leads to pretty accurate results. Finally, notice that correct results often look very different from the given examples, *i.e.*, QMAS is able to extrapolate from the given examples to other, correct tiles that do not have significant resemblance to the pre-labeled set.

6.4 Conclusions

In this chapter we described *QMAS*: *Q*uerying, *M*ining *A*nd *S*ummarizing *Multi-dimensional Databases*, one third algorithm that focuses on data mining in large sets of complex data. Specifically, *QMAS* is a fast solution to the problem of automatically analyzing, labeling and understanding this kind of data. Its main contributions, supported by experiments on real satellite images spanning up to 2.25 GB, are presented as follows:

1. **Speed**: *QMAS* is a fast solution to the problems presented, and it scales linearly on the database size. It is up to 40 times faster than top related works (GCap) on the same subject;
2. **Quality**: It can do *low-labor labeling* (*LLL*), providing results with accuracy better than or equal to the accuracy of the related works;
3. **Non-labor intensive**: It works even when it is given very few labels—*it can still extrapolate from tiny sets of pre-labeled data*;
4. **Functionality**: In contrast to the other methods, *QMAS* spots data objects that potentially require new labels, and encompasses other mining tasks such as clustering, outlier and representatives detection, as well as summarization;

The next chapter presents the conclusions of this book.

References

1. Bhattacharya, A., Ljosa, V., Pan, J.Y., Verardo, M.R., Yang, H.J., Faloutsos, C., Singh, A.K.: Vivo: visual vocabulary construction for mining biomedical images. In: ICDM, pp. 50–57. IEEE Computer Society (2005)
2. Cordeiro, R.L.F., Guo, F., Haverkamp, D.S., Horne, J.H., Hughes, E.K., Kim, G., Traina, A.J.M., Traina Jr., C., Faloutsos, C.: Qmas: querying, mining and summarization of multimodal databases. In: Webb, G.I., Liu, B., Zhang, C., Gunopulos, D., Wu, X. (eds.) ICDM, pp. 785–790. IEEE Computer Society (2010)
3. Cordeiro, R.L.F., Traina, A.J.M., Faloutsos, C., Traina Jr., C.: Finding clusters in subspaces of very large, multi-dimensional datasets. In: Li, F., Moro, M.M., Ghandeharizadeh, S., Haritsa, J.R., Weikum, G., Carey, M.J., Casati, F., Chang, E.Y., Manolescu, I., Mehrotra, S., Dayal, U., Tsotras, V.J. (eds.) ICDE, pp. 625–636. IEEE (2010)
4. Cordeiro, R.L.F., Traina, A.J.M., Faloutsos, C., Traina Jr., C.: Halite: fast and scalable multiresolution local-correlation clustering. IEEE Trans. Knowl. Data Eng. 99(PrePrints) (2011). doi:10.1109/TKDE.2011.176
5. Gibson, L., Lucas, D.: Spatial data processing using generalized balanced ternary. In: IEEE conference on pattern recognition and image analysis (1982)
6. Golub, G.H., Van Loan, C.F.: Matrix computations, 3rd edn. The Johns Hopkins University Press, Baltimore, USA (1996)
7. Huang, K., Murphy, R.F.: From quantitative microscopy to automated image understanding. J. Biomed. Optics **9**, 893–912 (2004)
8. Lloyd, S.: Least squares quantization in pcm. IEEE Trans Info Theory **28**(2), 129–137 (1982). doi:10.1109/TIT.1982.1056489

9. MacQueen, J.B.: Some methods for classification and analysis of multivariate observations. In: Cam, L.M.L., Neyman, J. (eds.) Proceedings of the fifth berkeley symposium on mathematical statistics and probability, vol. 1, pp. 281–297. University of California Press, California (1967)

10. Pan, J.Y., Balan, A.G.R., Xing, E.P., Traina, A.J.M., Faloutsos, C.: Automatic mining of fruit fly embryo images. KDD pp. 693–698 (2006)

11. Steinhaus, H.: Sur la division des corp materiels en parties. Bull. Acad. Polon. Sci. **1**, 801–804 (1956). (in French)

12. Zhang, B., Hsu, M., Dayal, U.: K-harmonic means—a spatial clustering algorithm with boosting. In: Roddick, J.F., Hornsby, K. (eds.) TSDM, lecture notes in computer science, vol. 2007, pp. 31–45. Springer, Heidelberg (2000)

Chapter 7
Conclusion

Abstract This book was motivated by the increasing amount and complexity of the dada collected by digital systems in several areas, which turns the task of knowledge discovery out to an essential step in businesses' strategic decisions. The mining techniques used in the process usually have high computational costs and force the analyst to make complex choices. The complexity stems from the diversity of tasks that may be used in the analysis and from the large amount of alternatives to execute each task. The most common data mining tasks include data classification, labeling and clustering, outlier detection and missing data prediction. The large computational cost comes from the need to explore several alternative solutions, in different combinations, to obtain the desired information. Although the same tasks applied to traditional data are also necessary for more complex data, such as images, graphs, audio and long texts, the complexity and the computational costs associated to handling large amounts of these complex data increase considerably, making the traditional techniques impractical. Therefore, especial data mining techniques for this kind of data need to be developed. We discussed new data mining techniques for large sets of complex data, especially for the clustering task tightly associated to other mining tasks that are performed together. Specifically, this book described in detail three novel data mining algorithms well-suited to analyze large sets of complex data: the method *Halite* for correlation clustering [11, 13]; the method *BoW* for clustering Terabyte-scale datasets [14]; and the method *QMAS* for labeling and summarization [12].

Keywords Big data · Complex data · Correlation clustering · Low-labor labeling · Summarization · Attention routing · Linear or Quasi-linear complexity · Terabyte-scale data analysis

R. L. F. Cordeiro et al., *Data Mining in Large Sets of Complex Data*,   111
SpringerBriefs in Computer Science, DOI: 10.1007/978-1-4471-4890-6_7,
© The Author(s) 2013

7.1 Main Contributions

Three data mining techniques were described in detail in this book. These techniques were evaluated on real, very large datasets with up to *billions* of complex elements, and they always presented highly accurate results, being at least one order of magnitude faster than the fastest related works in almost all cases. The real life datasets used come from the following applications: automatic breast cancer diagnosis, satellite imagery analysis, and graph mining on a large web graph crawled by Yahoo![1] and also on the graph with all users and their connections from the Twitter[2] social network. The three techniques are briefly discussed as follows.

1. **The Method** *Halite* **for Correlation Clustering**: the algorithm *Halite* [11, 13] is a fast and scalable density-based clustering algorithm for data of medium dimensionality able to analyze large collections of complex data elements. It creates a multi-dimensional grid all over the data space and counts the number of points lying at each hyper-cubic cell provided by the grid. A hyper-quad-tree-like structure, called the Counting-tree, is used to store the counts. The tree is thereafter submitted to a filtering process able to identify regions that are, in a statistical sense, denser than its neighboring regions regarding at least one dimension, which leads to the final clustering result. *Halite* is fast and has linear or quasi-linear time and space complexity on both data size and dimensionality.
2. **The Method** *BoW* **for Clustering Terabyte-scale Data**: the method *BoW* [14] focuses on clustering Terabytes of moderate-to-high dimensionality data, such as features extracted from billions of complex data elements. In these cases, a serial processing strategy is usually impractical. Just to read a single Terabyte of data (at 5GB/min on a single modern eSATA disk) one takes more than 3 hours. *BoW* explores parallelism through `MapReduce` and can treat as plug-in almost any of the serial clustering methods, including the algorithm *Halite*. The major research challenges addressed are (a) how to minimize the I/O cost, taking into account the *already existing* data partition (e.g., on disks), and (b) how to minimize the network cost among processing nodes. Either of them may be the bottleneck. *BoW* automatically spots the bottleneck and chooses a good strategy, one of them uses a novel *sampling-and-ignore* idea to reduce the network traffic.
3. **The Method** *QMAS* **for Labeling and Summarization**: *QMAS* [12] is a fast and scalable solution to the following problems: (a) **Low-labor labeling**, given a large set of complex objects, *very few* of which are labeled with keywords, find the most suitable labels for the remaining ones; and (b) **Mining and attention routing**, in the same setting, find clusters, the top-N_O outlier objects, and the top-N_R representative objects. The algorithm is fast and it scales linearly with the data size, besides working even with tiny initial label sets.

---

[1] www.yahoo.com

[2] twitter.com

7.2 Discussion

In the previous Chap. 3 we briefly describe representative methods from literature aimed at spotting clusters in moderate-to-high dimensionality data. Then, we conclude the chapter by summarizing in Table 3.1 relevant methods with regard to the main desirable properties that any clustering technique designed to analyze such kind of data should have. Here, we reprint in Table 7.1 the same table presented in Chap. 3, this time *including* the new methods described in the book. Notice that Table 7.1 was inspired[3] in one table found in [17].

Remember that the initial analysis of the literature provided in Chap. 3 lead us to come to one main conclusion. In spite of the several qualities found in the existing works, to the best of our knowledge, there is no method published in the literature, and designed to look for clusters in subspaces, that has *any* of the following desirable properties: (1) **to scale linearly or quasi-linearly** in terms of memory requirement and execution time with regard to increasing numbers of points and axes, besides increasing clusters' dimensionalities, and; (2) to be able to handle data of **Terabyte-scale** in feasible time.

One central goal of the work described in this book is to overcome these two limitations. Specifically, in Chap. 4, we focused on the former problem identified - **linear or quasi-linear complexity**. We described in detail the method *Halite*, a novel *correlation clustering* algorithm for multi-dimensional data, whose main strengths are that it is fast and it has linear or quasi-linear scalability in time and space with regard to increasing numbers of objects and axes, besides increasing clusters' dimensionalities. Therefore, the method *Halite* tackles the problem of **linear or quasi-linear complexity**. A theoretical study on the time and space complexity of *Halite*, discussed in Sect. 4.3, as well as an extensive experimental evaluation performed over synthetic and real data spanning up to 1 *million* elements and comparing *Halite* with seven representative works corroborate this claim.

In Chap. 5, we focused on the later problem - **Terabyte-scale data analysis**. We described in detail the method *BoW*, a novel, adaptive clustering method that explores parallelism using `MapReduce` for clustering huge datasets. It combines (a) potentially any serial algorithm used as a plug-in and (b) makes the plug-in run efficiently in parallel, by adaptively balancing the cost for disk accesses and network accesses, which allows *BoW* to achieve a very good tradeoff between these two possible bottlenecks. Therefore, *BoW* tackles the problem of **Terabyte-scale data analysis**. Experiments performed on both real and synthetic data with *billions* of points, and using up to 1, 024 cores in parallel corroborate this claim.

Finally, notice in Table 7.1 that the use of the *Halite* algorithm as a plug-in for the *BoW* method creates one powerful tool for clustering moderate-to-high dimensionality data of large scale-this configuration has both the desirable properties

---

[3] Table 7.1 includes a summary of one table found in [17], i.e., Table 7.1 includes a selection of most relevant desirable properties and most closely related works from the original table. Table 7.1 also includes two novel desirable properties not found in [17]—**Linear or quasi-linear complexity** and **Terabyte-scale data analysis**.

Table 7.1 Properties of methods aimed at clustering moderate-to-high dimensionality data, *including* the new methods *Halite* and *BoW* (using *Halite* as plug-in)

Clustering Algorithm	Arbitrarily oriented clusters	Not relying on the *locality assumption*	Adaptive density threshold	Independent of the order of the attributes	Independent of the order of the objects	Deterministic	Arbitrary number of clusters	Overlapping clusters (soft clustering)	Arbitrary subspace dimensionality	Avoiding complete enumeration	Robust to noise	Linear or quasi-linear complexity	Terabyte-scale data analysis
Axes parallel clustering													
CLIQUE [6, 7]		✓		✓	✓	✓	✓	✓	✓			✓	
ENCLUS [10]		✓		✓	✓	✓	✓	✓	✓			✓	
SUBCLU [18]		✓		✓	✓	✓	✓	✓	✓			✓	
PROCLUS [5]			✓							✓			
PreDeCon [8]				✓	✓	✓	✓				✓	✓	
P3C [19, 20]		✓	✓	✓	✓	✓	✓	✓	✓			✓	
COSA [15]				✓	✓	✓				✓	✓	✓	
DOC/FASTDOC [21]		✓		✓	✓		✓	✓	✓				
FIRES [16]		✓	✓		✓	✓	✓	✓	✓	✓	✓	✓	
Correlation clustering													
ORCLUS [3, 4]	✓				✓					✓			
4C [9]	✓			✓	✓	✓	✓				✓	✓	
COPAC [2]	✓			✓	✓	✓	✓			✓	✓	✓	
CASH [1]	✓	✓	*n a*	✓	✓	✓				✓		✓	
Halite [11, 13]	✓			✓	✓	✓	✓	✓	✓	✓	✓	✓	
BoW [14]	✓			✓	✓	✓	✓	✓		✓	✓	✓	✓

Notice that this table was inspired in one table found in [17]. *n a*: not applicable

sought, **linear or quasi-linear complexity** and **Terabyte-scale data analysis**, still having most of the other properties that any clustering technique designed to analyze moderate-to-high dimensionality data should have.

In summary, the work described in this book takes steps forward from traditional data mining (especially for clustering) by considering large, complex datasets. Note that, usually, current works focus in one aspect, either data size or complexity. The work described here considers both: it enables mining complex data from high impact applications, such as breast cancer diagnosis, region classification in satellite images, assistance to climate change forecast, recommendation systems for the Web and social networks; the data are large in the Terabyte-scale, not in Giga as usual; and very accurate results are found in just minutes. Thus, it provides a crucial and well timed contribution for allowing the creation of *real time* applications that deal with *Big Data of high complexity* in which mining on the fly can make an immeasurable difference, like to support cancer diagnosis or deforestation detection.

References

1. Achtert, E., Böhm, C., David, J., Kröger, P., Zimek, A.: Global correlation clustering based on the hough transform. Stat. Anal. Data Min **1**, 111–127 (2008). doi:10.1002/sam.v1:3
2. Achtert, E., Böhm, C., Kriegel, H.P., Kröger, P., Zimek, A.: Robust, complete, and efficient correlation clustering. SDM, USA (2007)
3. Aggarwal, C.C., Yu, P.S.: Finding generalized projected clusters in high dimensional spaces. SIGMOD Rec. **29**(2), 70–81 (2000). doi:10.1145/335191.335383
4. Aggarwal, C., Yu, P.: Redefining clustering for high-dimensional applications. IEEE TKDE **14**(2), 210–225 (2002). doi:10.1109/69.991713
5. Aggarwal, C.C., Wolf, J.L., Yu, P.S., Procopiuc, C., Park, J.S.: Fast algorithms for projected clustering. SIGMOD Rec. **28**(2), 61–72 (1999). doi:10.1145/304181.304188
6. Agrawal, R., Gehrke, J., Gunopulos, D., Raghavan, P.: Automatic subspace clustering of high dimensional data for data mining applications. SIGMOD Rec. **27**(2), 94–105 (1998). doi:10.1145/276305.276314
7. Agrawal, R., Gehrke, J., Gunopulos, D., Raghavan, P.: Automatic subspace clustering of high dimensional data. Data Min. Knowl. Discov. **11**(1), 5–33 (2005). doi:10.1007/s10618-005-1396-1
8. Bohm, C., Kailing, K., Kriegel, H.P., Kroger, P.: Density connected clustering with local subspace preferences. In: ICDM '04: Proceedings of the Fourth IEEE International Conference on Data Mining, pp. 27–34. IEEE Computer Society, USA (2004).
9. Böhm, C., Kailing, K., Kröger, P., Zimek, A.: Computing clusters of correlation connected objects. In: SIGMOD, pp. 455–466. USA (2004). http://doi.acm.org/10.1145/1007568.1007620
10. Cheng, C.H., Fu, A.W., Zhang, Y.: Entropy-based subspace clustering for mining numerical data. In: KDD, pp. 84–93. NY, USA (1999). http://doi.acm.org/10.1145/312129.312199
11. Cordeiro, R.L.F., Traina, A.J.M., Faloutsos, C., Traina Jr, C.: Finding clusters in subspaces of very large, multi-dimensional datasets. In: Li, F., Moro, M.M., Ghandeharizadeh, S., Haritsa, J.R., Weikum, G., Carey, M.J., Casati, F., Chang, E.Y., Manolescu, I., Mehrotra, S., Dayal, U., Tsotras, V.J. (eds.) pp. 625–636. IEEE In ICDE. (2010).
12. Cordeiro, R.L.F., Guo, F., Haverkamp, D.S., Horne, J.H., Hughes, E.K., Kim, G., Traina, A.J.M., Traina Jr., C., Faloutsos, C.: Qmas: Querying, mining and summarization of multimodal databases. In: Webb, G.I., Liu, B., Zhang, C., Gunopulos, D., Wu, X. (eds.) ICDM, pp. 785–790. IEEE Computer Society (2010).
13. Cordeiro, R.L.F., Traina, A.J.M., Faloutsos, C., Traina Jr., C.: Halite: Fast and scalable multiresolution local-correlation clustering. IEEE Trans. Knowl. Data Eng. 99(PrePrints) (2011). doi:10.1109/TKDE.2011.176.
14. Cordeiro, R.L.F., Traina Jr., C., Traina, A.J.M., López, J., Kang, U., Faloutsos, C.: Clustering very large multi-dimensional datasets with mapreduce. In: C. Apté, J. Ghosh, P. Smyth (eds.) KDD, pp. 690–698. ACM (2011).
15. Friedman, J.H., Meulman, J.J.: Clustering objects on subsets of attributes (with discussion). J. Roy. Stat. Soc. Ser. B **66**(4), 815–849 (2004). doi:a/bla/jorssb/v66y2004i4p815-849.html
16. Kriegel, H.P., Kröger, P., Renz, M., Wurst, S.: A generic framework for efficient subspace clustering of high-dimensional data. In: ICDM, pp. 250–257. Washington, USA (2005). http://dx.doi.org/10.1109/ICDM.2005.5

17. Kriegel, H.P., Kröger, P., Zimek, A.: Clustering high-dimensional data: A survey on subspace clustering, pattern-based clustering, and correlation clustering. ACM TKDD **3**(1), 1–58 (2009). doi:10.1145/1497577.1497578
18. Kröger, P., Kriegel, H.P., Kailing, K.: Density-connected subspace clustering for high-dimensional data. SDM, USA (2004)
19. Moise, G., Sander, J., Ester, M.: P3C: A robust projected clustering algorithm. In: ICDM, pp. 414–425. IEEE Computer Society (2006).
20. Moise, G., Sander, J., Ester, M.: Robust projected clustering. Knowl. Inf. Syst. **14**(3), 273–298 (2008). doi:10.1007/s10115-007-0090-6
21. Procopiuc, C.M., Jones, M., Agarwal, P.K., Murali, T.M.: A monte carlo algorithm for fast projective clustering. In: SIGMOD, pp. 418–427. USA (2002). http://doi.acm.org/10.1145/564691.564739